HAYFA BERGAOUI
Adam Ben Nasr
Amel Bouzouita

New blood culture identification techniques

HAYFA BERGAOUI
Adam Ben Nasr
Amel Bouzouita

New blood culture identification techniques

ScienciaScripts

Imprint
Any brand names and product names mentioned in this book are subject to trademark, brand or patent protection and are trademarks or registered trademarks of their respective holders. The use of brand names, product names, common names, trade names, product descriptions etc. even without a particular marking in this work is in no way to be construed to mean that such names may be regarded as unrestricted in respect of trademark and brand protection legislation and could thus be used by anyone.

Cover image: www.ingimage.com

This book is a translation from the original published under ISBN 978-620-6-70902-2.

Publisher:
Sciencia Scripts
is a trademark of
Dodo Books Indian Ocean Ltd. and OmniScriptum S.R.L publishing group

120 High Road, East Finchley, London, N2 9ED, United Kingdom
Str. Armeneasca 28/1, office 1, Chisinau MD-2012, Republic of Moldova, Europe
Printed at: see last page
ISBN: 978-620-8-22979-5

Contents

INTRODUCTION

Sepsis is defined as "life-threatening organ dysfunction caused by a disproportionate host response to infection". Although the treatment of sepsis has developed rapidly in recent years, the incidence and mortality continue to rise. Sepsis remains the leading cause of death in intensive care units, affecting more than 30 million people worldwide each year and is one of the leading causes of death worldwide. Any infected person can potentially develop sepsis. The incidence of sepsis reaches 1 to 2% of all hospitalised patients, resulting in prolonged hospital stays in specialised facilities such as intensive care units, and significant additional expenditure. American data estimates the incidence of sepsis at around 300 cases/100,000 people, and the mortality rate varies between 30% and 50%, more than the number of deaths caused by prostate cancer, breast cancer and AIDS combined. In addition, because of the diverse manifestations of sepsis, clinicians continue to face serious challenges in diagnosis and management. The cost of treatment is very high: in 2013, the total cost was approximately 23.6 billion dollars for almost 1.3 million hospitalised patients in the United States, and this has been increasing at an annual rate of 11.5%. Mortality increases by 7% for each hour of delay in administering appropriate antibiotic therapy. Moreover, even after treatment, patients face serious long-term problems in the form of physical, psychological and cognitive disorders. In 2012, the Global Sepsis Alliance and its founding members established World Sepsis Day on 13 September to raise public awareness, and there is an urgent need to develop rapid and effective methods for the prevention, diagnosis and treatment of sepsis**[1-3]**.

The reference technique for detecting the microorganisms responsible is blood culture. However, there are still some major drawbacks: an excessively long turnaround time, an inability to detect slow-growing intracellular micro-organisms and pathogenic agents other than bacteria or yeasts. These drawbacks can have far-reaching consequences. As culture-based diagnosis is slow, doctors often have to start empirical broad-spectrum antibiotic treatment, and starting treatment with an appropriate antibiotic as soon as possible, preferably within the first hour of clinical recognition of sepsis, is crucial for reducing morbidity and mortality. Consequently, rapid detection and identification of pathogens is necessary to optimise antibiotic treatment and de-escalation in order to limit side-effects and the selection of antibiotic resistance. In addition, the long turnaround time can also lead to delayed specific clinical diagnosis, increased length of stay, higher complication rates and ultimately increased costs **[4]**.Therefore, for more than a decade, specific tests aimed at the early identification of pathogens and certain markers of resistance have been developed and tested.

The aim of this study is to provide an update on new techniques for identifying micro-organisms and testing antibiotic susceptibility from positive blood culture bottles, in order to optimise patient management.

CHAPTER 1

1. HEMOCULTURE

Blood is normally a sterile medium. A bacteremia corresponds to the presence of bacteria in the blood, and is generally associated with a sign of infection such as high fever, which may be accompanied by chills and sweating and, in rare cases, hypothermia. It is generally related to a focal infectious site (pneumonia) or a translocation of digestive flora or an endovascular site (endocarditis)**[5]**. Blood culture is used when a bacterial or fungal blood infection is suspected, in order to identify the microorganisms responsible and ensure adequate antibiotic therapy**[6]**.

1.1. The pre-analytical phase

The collection of biological samples is a crucial pre-analytical activity.errors occurring shortly before or after this stage can alter the quality of the samples or modify the final results.checking the sampling techniques and procedures as well as the equipment used to collect the blood can prevent many problems**[7].**

1.2. Blood culture sampling protocol

Blood sampling should preferably be carried out before any antibiotic therapy is started, and essentially involves collecting venous blood directly from the vein or from a central or peripheral catheter and inoculating it directly into liquid culture media contained in vials, with a lower risk of contamination than with venipuncture**[8]**. Strict aseptic measures must be applied by healthcare staff: washing hands with a hydroalcoholic solution, wearing gloves and skin antisepsis of the puncture area and disinfection of the vial lid with an alcoholic chlorhexidine solution for greater efficacy**[8]**.

Table I **lists** the key elements for optimising blood cultures that may have an impact on contamination.

Table I: essential elements for reducing contamination and optimising blood cultures [9].

Element	How do we do it?
Body site for sample collection	Peripheral venipuncture preferred
Preparing the venipuncture site	Use of alcoholic chlorhexidine gluconate and/or tincture of iodine with 70% alcohol. Iodophores (povidone iodine) are not recommended.
Taking samples **Hand hygiene** **Sterile gloves** **Number of bottles** **Volume of withdrawals** **Transporting bottles**	Using a team of trained carers Compliance with aseptic practices A minimum of 2 samples, each sample combining an aerobic and anaerobic bottle 8 to 10 mL per vial, 20 to 30 mL per venipuncture As quickly as possible, at room temperature
Use of a permanently monitored blood culture system Rapid performance of Gram staining Rapid	Reduces the lead time Processing of positive samples 24/7 Allows faster identification of contaminants and pathogens

1.3. Number and volume of bottles

An optimal filling volume and a sufficient number of vials determine diagnostic sensitivity. Each sample requires 2 blood culture bottles to be filled. The first in aerobiosis and the second in anaerobiosis, to allow the growth of germs preferring one or other of the components, it is also advisable to perform 2 to 3 blood culture series (i.e. 4 to 6 vials). Two series are sufficient, since 80-90% of bacteria are detected in the first sample, and 88-99% in the 2[eme] sample. The samples may be taken simultaneously (at 1 or 2 different sites) or spaced out over time (at least 30 min) **[5]**.

A minimum volume of 10 mL per vial should be drawn for adults and 1-3 mL for children. Gaur et al. suggested a volume of blood depending on the weight of the child and infant: 1 mL for children weighing less than 2kg, 1.5 mL for patients weighing less than 11kg and 7.5mL for patients between 11kg and 17kg and 16.5 mL for children weighing >37kg **[10]**.

Commercially available blood culture bottles are formed from a liquid broth (trypticase soya, for example) with the addition of CO_2 and the nutrients required for bacterial growth (cysteine, hemin, sugars, etc.) Blood culture bottles contain an anticoagulant, often sodium polyanethol sulphonate (SDS) (at a concentration ranging from 0.025% to 0.05%), which neutralises certain antibiotics**[11]**.

1.4. Transporting vials to the laboratory

The time it takes to transport blood culture bottles to the laboratory and incubate them in the automated systems is a critical factor. Any delay can delay or hinder bacterial growth. The acceptable limit is 4 hours. Beyond this time, a delay could adversely affect the rate of positive blood cultures and the time to rapid and effective antibiotic treatment. Inoculated vials, if not taken immediately to the laboratory and incubated, should be stored at room temperature, avoiding temperatures above 30°C or refrigerated conditions**[12]**.

1.5. Incubation of vials

In most modern microbiology laboratories, vials are incubated in automated systems. These systems are formed by a series of individual cells with continuous agitation to facilitate bacterial multiplication. They detect the growth of microorganisms by measuring the carbon dioxide released either by reflectometry (BACT/ALERT®), fluometry (BD BACTEC™) or by detecting pressure changes in the top of a closed vial (versaTREK™). An audible and/or visual alarm warns of any positive result **[6]**.

A total of 85-90% of positive bottles are detected in less than 48 hours, except in the case of fungemia or bacteremia caused by a slow-growing species. Vials are usually incubated for 5 days before being reported as negative. This period should be extended in certain situations such as suspected endocarditis **[13]**.

CHAPTER 2

2. WHAT TO DO IF A BOTTLE DETECTS POSITIVE

2.1. Direct examination

Microscopic examinations are used to determine whether the blood culture appears to be mono-microbial or multi-microbial. A fresh state is performed to observe the mobility and morphology of the bacteria, and a Gram stain is performed, the result of which is immediately communicated to the clinician to adjust or initiate anti-biotherapy**[14]**.Gram staining makes it possible to distinguish between Gram-positive cocci in chains (Staphylococcus dore, *Staphylococcus* a coagulase negative (SCN)...), Gram-positive cocci in clusters (Streptococci, *Enterococcus faecalis*, *Enterococcus faecium.*Gram-positive diplococci (*Streptococcus pneumoniae...*), Gram-negative bacilli (*Enterobacteriaceae, Pseudomonas aeruginosa, Bacteroides fragilis* group, fuso bacteria...), Gram-positive bacilli (*Listeria monocytogenes*, *Propionibacterium acnes, Corynbacterium spp.*), Gram-negative diplococci (*Neisseria meningitidis.*), yeasts (*Candida albicans*, candida non albicans.)**[15]**.

2.2. Classic analysis techniques

2.2.1 Classic subculture

Depending on the results of the direct examination, inoculation on non-selective and selective gelose culture media adapted to the type of bacteria observed should be performed in order to carry out an identification and an antibiogram **[15]**.

2.2.2 Identification and antibiotic susceptibility testing

A series of specific biochemical tests for each group of microorganisms were carried out on colonies obtained from the seeded media. Gram-positive cocci are subjected to the catalase test to differentiate staphylococci from enterococci, and a series of tests are performed to identify species of the *Staphylococcus* genus: coagulase, fermentation of sugars (sucrose, maltose, trehalose, xylose, mannitol, etc.). Isolates identified as Gram-positive, catalase-negative, bile-esculin-positive, NaCl-positive and cocci positive to the pyrrolidonyl aminopeptidase test were subjected to biochemical tests for fermentation of mannitol, arabinose, arginine and sorbitol, and the presence or absence of a pigment on sheep's blood gelose. Bacilli fermenting glucose are subjected to the following biochemical tests: H_2S production, urease and L-tryptophan desaminase, mobility, indole production, lysine decarboxylase production and the ability to use citrate as a sole carbon source... Gram-negative bacilli that do not ferment glucose are identified on the basis of mobility, growth at a temperature of 42°C, DNAse production, etc. Yeasts are isolated on Sabouraud gelose and identified on the basis of colour, texture, colony shape and biochemical characteristics **[16]**.

The diffusion method is one of the oldest approaches for determining the sensitivity of bacteria to antibiotics, and remains one of the most widely used routine methods. It is suitable for the majority of pathogenic bacteria, including slow-growing bacteria. It allows a variable choice of antibiotics and requires no special equipment**[17]**.

A broth dilution test is often useful for determining the minimum inhibitory concentration (MIC). The MIC is determined by performing a sequence of decreasing concentrations of the drug in broth, which is then inoculated with the bacteria being tested. After incubation, turbidity can be read visually, although clinical laboratories with high workloads can read plates with spectrophotometers **[18]**.

2.3. Rapid analysis techniques for positive blood culture bottles 2.3.1 *Matrix-assisted laser desorption ionization time-of-flight*

The basic principle of *Matrix-Assisted Laser Desorption Ionization Time-Of-Flight* (MALDI-TOF) is **illustrated in Figure 1**.

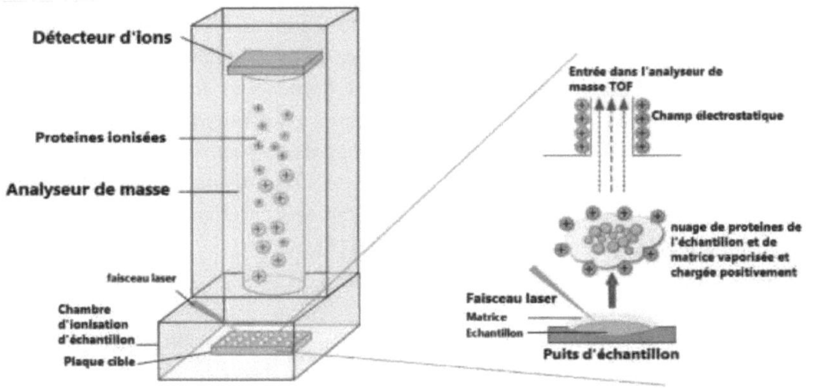

Figure 1: MALDI-TOF mass spectrometer [19].

In brief, bacterial samples deposited on a target plate are coated with a matrix solution (alpha cyano-4-hydroxycinnamic acid (**ACHC**) reconstituted in a mixture generally consisting of water and the organic solvents acetonirile and trifluroacetic acid) which co-crystallises with the sample and lyses the micro-organisms. Bacteria that are more difficult to lyse, such as certain Gram-positive bacteria, mycobacteria, yeasts and moulds, often require additional pre-treatment with a strong organic acid or mechanical lysis. Once placed in the instrument, samples are ionised by short laser pulses, forming gas-phase ions with minimal fragmentation. The time it takes for each particle to reach the detector, or the 'time of flight', depends on its mass and charge. Once all the abundant proteins in the sample have been detected by the mass spectrometer, a spectral fingerprint is produced that is unique to the organism being analysed. Studies have shown that the predominant proteins detected by MALDI-TOF are ribosomal proteins, although other highly abundant cytosolic proteins are also detected. The identification of the test organism is then determined automatically using software that compares the spectral profile of the unknown organism with a reference database **[19]**. A score < 1.7 indicates no reliable identification, a score between 1.7 and 1.999 indicates identification at genus level and a score > 2 indicates identification at species level **[20]**.

The two MALDI-TOF systems most commonly used in laboratories for microbial identification (**Figure 2**) are the Bruker MALDI Biotyper (Bruker Daltonics, Billerica, MA) and the bioMerieux VITEK MS (bioMerieux, Durham, NC) **[21]**.

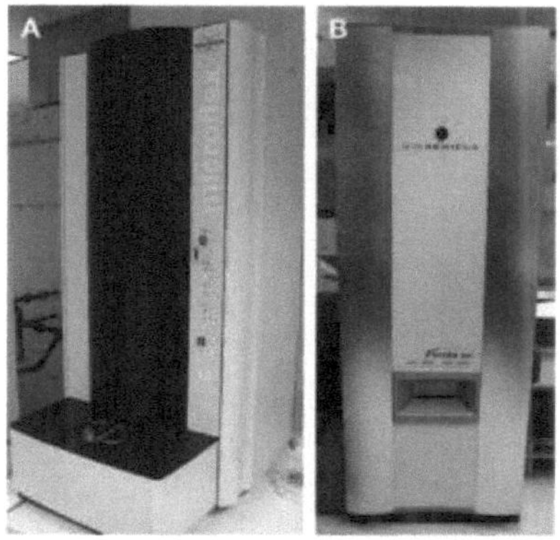

Figure 2: (A) Bruker Daltonics MALDI Biotyper (B) bioMerieux VITEK MS [21].

MALDI-TOF mass spectrometry (MS) is one of the most widely used methods for the identification of microorganisms in clinical samples, is one of the most promising technologies currently available for the identification of microbial pathogens directly from positive blood culture bottles and has considerable importance as a rapid diagnostic method, achieving a positive predictive value of 60-80% for different species of microorganisms.

Differentiation of microorganisms from host cells is a critical step for successful identification, and several laboratory-developed and commercially available protocols have been reported for this purpose.Several proteins may be present in large quantities that do not originate from bacteria, such as haemoglobin. Therefore, pre-treatment steps, such as blood cell separation, are required to selectively recover bacterial cells.Tsuchida et al. have studied the functionality of various methods and kits. Among these, the Sepsityper ® MALDI kit (Bruker Daltonics) is widely used, which employs a lysis and centrifugation strategy: this involves mixing 200p,L of the kit's lysis buffer with 1mL of blood culture to break down the blood components of the sample, the bacterial cells are then collected by high-speed centrifugation**[22]**. **Table II** describes the Sepsityper method used with the MALDI Biotyper.

Another method is to mix 2mL of blood culture and 1mL of lysis buffer to break down the blood components, then collect the bacteria using a 0.45 ^m *Millipore Express PLUS* filter. The VITEK® MS kit (bioMerieux) uses a lysis and suction filtration strategy, collecting bacteria through a filter, the bacteria collection device uses a glass collector and vacuum pump. Once the bacteria have been collected, the plate method is used for extraction.

Using tailor-made methods, proteins from red blood cells and other host cells in the blood culture are selectively removed and bacteria can be collected efficiently. A method that selectively removes haemoglobin has recently been developed. The negative charge on the surface of the micro-organisms is used to capture the micro-organisms by attraction to the positive charge of the polyallylamine (cationic granules). This product has been upgraded to a convenient pre-treatment kit manufactured in Japan. It is sold in Japan and abroad under the name rapid BACpro® II and is currently undergoing clinical evaluation**[22]**.

Table II: Steps used in sample preparation by Sepsityper and results obtained

Preparation of the sample	Results	Reference
Mix 1 mL of the blood culture bottle with 200 pL of lysis buffer. Shake for 10 s. Centrifuge for 2 min at 13,000 rpm Wash the pellet in 1 mL of wash buffer. Centrifuge for 1 min at 13,000 rpm Resuspend the pellet in 300 pL of HPLC quality water (High Performance Liquid Chromatography) and add 900 pL of ethanol. Centrifuge for 2 min at 13,000 rpm. Dry the pellet at ambient temperature Resuspend in 250 pL of funonic acid and acetonitrile. Centrifuge for 2 min, 13,000 rpm Transfer 1 pL of the pellet to the target plate for MALDI-TOF analysis.	For Gram-positive bacteria: **90%** identifiable, **57%** identified at genus level, **33%** identified at species level. For Gram-negative bacteria**: 92%** identifiable, **42%** identified at genus level, **51%** identified at species level. For anaerobic bacteria: **67%** identifiable, **58%** identified at genus level, **8%** identified at species level.	**[20]**
Mix 1 mL of blood culture bottles with lysis buffer (100 pL of 20% sodium dodecyl sulphate (DSS) solution and 200 pL of 5% saponin solution). Vortex for 10 s. Centrifuge for 2 min, 16,000 g Wash the pellet twice with 1 mL distilled water. Discard the supernatant **For the rapid extraction (RE) protocol:** place 1 pL of the wet pellet in duplicate on the target plate before adding 1 pL of funonic acid and 1 pL of ACHC. **For the long extraction protocol (EL):** resuspend the remaining pellet in 300 pL of distilled water. Apply the ethanol/fungal acid procedure. Place 1 pL of the duplicate sumageant on the target plate and add the ACHC.	For Gram-negative bacteria, identification is **100%, 100%, 75%** and **90%** for DSS-EL, DSS-ER, SAP-EL and SAP-ER respectively. For Gram-positive bacteria, identification was **82%, 23%, 23%** and **0%** for DSS-EL, DSS-ER, Saponin-EL and Saponin-ER.	**[23]**
Mix 1 mL of blood culture with 200 pL of lysis buffer in an Eppendorf tube. Mix for 10 s. Centrifuge for 1 min at 3000 rpm. Suspend the pellet in 1 mL wash buffer. Centrifuge for 1 min at 13,000 rpm. Place 1 pL of supernatant and 1 pL of matrix solution on target slides.	Gender identification only was obtained in **12.4%** of cases.	**[24]**
Mix 1 mL of blood culture with 200 pL lysis buffer. Vortex for 10 s. Centrifuge for 2 min at 13000 rpm. Suspend the pellet with 1 mL of wash buffer. Mix. Centrifuge for 1 min at 13000 rpm Resuspend the pellet with a 70% ethanol solution. Vortex. Centrifuge and remove the sumageant.	Anaerobies **70.6%** correctly identified Gram-negative Anaerobia **46.5%** correctly identified Gram-positive Anaerobia **70.8%** correctly identified	**[25]**

Resuspend the pellet in 30 pL of 70% funonic acid and an equal quantity of 100% acetonitrile. Mixing and centrifuging Place 1 pL of supernatant on the target plate and air dry + 1 pL of ACHC matrix solution and analyse with MALDI-TOF.		
Mix 1 mL of blood culture broth in an Eppendorf tube with 200 pL of lysis buffer. Centrifuge for 2min, 16 060g Add 1mL of wash buffer to the pellet. Centrifuge at 16 060g. Remove the supernatant by pipetting Resuspend the pellet by adding 300 pL of HPLC-grade water. Add 900 pL ethanol. Centrifuge 2 min, 16 060 g Resuspend the pellet in 20 pL -50 pL of 70% funonic acid and the same amount of 100% acetonitrile. Centrifuge Imin at 16060g. Place 1 pL of supernatant on the target plate and allow to dry. Add the matrix solution.	Gram-positive bacteria **62.6%** reliable identification Gram-negative bacteria; **76.2%** reliable identification	**[26]**
Mix 1 mL of hemoculture broth with 200 pL of lysis buffer. Vortex for 10 sec. Centrifuge for 1 min Resuspend the pellet in 1 mL of wash buffer. Centrifuge. Resuspend the pellet with 300 pL of distilled water.	Gram-positive bacteria; **76.1%** correct identification Gram-negative bacteria; **89.6%** correct identification	**[27]**
Add 900 pL ethanol. Centrifuge for 2 min Remove the residual ethanol. Resuspend the pellet with 30 pL formic acid and 30 pL acetonitrile. Vortex. Briefly centrifuge Apply IpL of 1 extract to the target plate and allow to dry.	Yeast: **65.9%** correct identification	
Mix 1 mL of hemoculture broth with 200 pL of lysis buffer. Mix well. Centrifuge for 2 min at 13000 rpm. Vortex for 10 s. Discard the supernatant Resuspend the pellet with 1 mL of wash buffer. Centrifuge Imin at 13000 rpm. Remove the sumageant. Add 300 pL of demineralised water and 900 pL of ethanol Centrifuge for 2 min at 13,000 rpm. Discard supernatant. Centrifuge and dry. Add 50 pL of 70% formic acid and 50 pL of acetonitrile. Mix and centrifuge for 2 min at 13000 rpm. Take 4 volumes of 1 pL of the supernatant and place on the target plate and dry. Add 1 pL of matrix solution and dry.	**For polymicrobial vials:** Gram-negative bacteria; **61.8%** identified Gram-positive bacteria; **59.9%** identified Two correct identifications occurred in **45.5%** of cases. A single micro-organism was identified in **50% of** cases and no identification was made in **4.5%** of cases.	**[28]**

Mix 1 mL of blood culture broth with 200 pL of lysis reagent. Vortex 10 s and leave to stand for 5 min at room temperature. Centrifuge for 1 min at 13,000 g. Discard the supernatant and return the pellet with 1 mL of wash buffer. Centrifuge for 1 min at 13,000 g. Discard the supernatant and return the pellet with 300 pL of quality water and 900 pL of 100% ethanol. Centrifuge for 2 min at 13000 g. Decant the supernatant. Centrifuge for 2 min at 13000g. Remove all residual ethanol and dry the pellet for 10 min at room temperature. Add 30 pL of 70% formic acid and 30 pL of 100% acetonitrile. Centrifuge for 2 min at 13000 g Add 1 pL of the supernatant to the target plate and dry. Add 1 pL of matrix solution and dry.	<u>For Gram-positive bacteria</u>: **84.7%** reliable identification **56.8%** identified to species level. <u>For Gram-negative bacteria;</u> **85.2%** identified at species level.	**[29]**
Mix 1 mL of blood culture broth with 200 pL of lysis buffer.	<u>For Gram-positive bacteria</u>:	**[30]**
Vortex 10 s. Centrifuge for 1 min at 13,000 rpm. Discard the supernatant and return the pellet with 1 mL of wash buffer. Vortex 10 s and centrifuge Imin at 13000 rpm. Discard the sumageant and extract with ethanol and funonic acid. Add 1 pL of sumageant to the target plate and dry. Add 1 pL of matrix solution and dry.	69.6% identified by gender, 35.9% identified at species level. Most misidentifications occurred with *Staphylococcus epidermidis* <u>For Gram-negative bacteria;</u> 97.2% identified by gender, 76.4% identified at species level. <u>Yeast:</u> 72.7% identifies an gender	
Add 200 pL of each reagent (5% saponin solution, 10% DSS solution, Bruker SepsiTyper lysis buffer) to 3 different tubes + 1 mL of blood culture broth. Mix and vortex for 1 s Incubate for 5 min at room temperature. Centrifuge for 1 min at 13000g. Remove the sumageant and resuspend the pellet with 1 mL of HPLC-grade water for the 2 saponin and DSS tubes and with 1 mL of wash buffer for the Sepsityper tube. Centrifuge for 1 min at 13000 g. Remove the supernatant and resuspend the pellet with 300 pL of quality water and 900 pL of ethanol. Centrifuge for 2 min at 13000 g. Discard the supernatant and centrifuge for a further 2 min at 13000 g. Remove the residual ethanol and dry the pellet for 10 min at room temperature. Add one volume of 70% funonic acid and an equal volume of acetonitrile to the pellet.	<u>**For polymicrobial vials**</u>: <u>Saponin :</u> A single organisation identified 63% of reliable identification and 34% at species level. <u>DSS :</u> A single organism was identified in 66% of cases <u>SepsiTyper ;</u> A single organisation was identified in 69% of cases. <u>**For the single-microbe vial**</u> **:** The lowest yield of species identification was obtained with saponin and the highest yield was obtained using SepsiTyper,	**[31]**

Centrifuge for 2 min at 13000 g Add 1 pL of each supernatant to the target plate and dry. Add 1 pL of matrix solution and dry.		

rpm: revolutions per minute; HPLC: high performance liquid chromatography; DSS: sodium dodecyl sulphate; ACHC: alpha-cyano-4-hydroxycinnamic acid.

Table III: Steps used to prepare the sample using in-house methods and results obtained

Sample preparation method	MALDI-TOF equipment	Results	Reference
Saponin method Mix 1 mL of blood culture broth with 500 pL of 4% saponin. Vortex for 10s. Incubate for 3 min at room temperature and centrifuge for 1 min at 13000 rpm. Discard the sumageant and resuspend the pellet with 1 mL of distilled water. Vortex for 10s and centrifuge for 1 min, 13000 rpm. Discard the sumageant and extract with ethanol and funonic acid. Add 1 pL of sumageant to the target plate and dry. Add IpL of matrix solution and dry.	MALDI Biotyper	For Gram-positive bacteria: **69.6%** identified at genus level. **29.3%** identified to species level. For Gram-positive bacilli**: 40%** identified at genus level. For Gram-negative bacilli**:** **94.4%** identified at genus level, **72.2%** identified at species level. Yeasts **81.8%** identified by gender.	**[30]**
Saponin method Mix 1 mL of blood culture volume with 200 pL of 5% saponin. Vortex and incubate for 5 minutes at room temperature. Centrifuge for 1 minute at 14500 rpm. Wash the pellet with 1 mL deionised water. Centrifuge Imin, 14500 rpm. Place 1 pL of supernatant on the target plate and air dry + 1 pL of matrix solution (ACHC)	MALDI Biotyper	For Gram-positive anaerobes**: 62.5%** reliable identification. For Gram-negative anaerobes; **100%** reliable identification.	**[25]**
Saponin method Mix 1 mL blood culture and 200 pL lysis buffer with 2% saponin. Shake for 30s. Leave at room temperature for 5 min. Centrifuge for 1 min, 14,500g. Homogenise the pellet with 1 mL sterile distilled water. Shake until completely dissolved. Centrifuge for 1 min, 14,500g. Wash the pellet with sterile distilled water, repeat 2 times. Homogenise the spots from the pellet and dry at room temperature.	bioMerieux VITEK® MS	Gram-positive bacteria; **78.5%** correct identification. Gram-negative bacteria; **93.5%** correct identification.	**[32]**
ambient. Cover each well with 0.5 pL of formic acid and 1 pL of matrix solution.			
Mix 1 mL of blood culture with 200 pL of 5% saponin lysis solution. Vortex for 10 s Centrifuge for 60 s, 15400g. Wash the pellet several times by pipetting with 1 mL deionised	MALDI Biotyper	For Gram-positive bacteria: **93.4%** identified at genus level. **83.6%** identified at species level. For Gram-	**[33]**

water. Centrifuge for 60 s, 15400g. Subject the pellet to extraction with 50pL of 70% formic acid. After 30 s of incubation, add 50 pL of acetonitrile Centrifuge 60 s, 15400g Apply 1 pL to the dry target plate and add 1 pL of ACHC matrix solution and dry.		negative bacteria: **89%** identified at genus level. **83.5%** identified at species level. Yeasts : **62.5%** identifies at gender level. **50%** identifies at species level.	
Transfer 1 mL of broth to a tube (A) and 500 pL of lysis buffer. Centrifuge for 3 min, 2000 g. Discard the supernatant and add 800 pL of distilled water Add 200 pL of reaction buffer and 200 pL of a cationic copolymer to a 2[eme] tube (B). Vortex. Transfer tube A to tube B. Vortex. Centrifuge for 1 min, 2000 g. Discard the sumageant and add 800 pL of 70% ethanol. Centrifuge for 1 min, 2000 g Discard the supernatant and add 30 pL formic acid and 30 pL acetonitrile. Vortex and centrifuge for 1 min, 2000 g. Add 1 pL of the supernatant to the target plate and dry.	bioMerieux VITEK® MS	**96.3%** correct identification.	**[34]**
Transfer 3 mL of the blood culture broth into a 3.5 mL tube containing plasma separation gel. Centrifuge for 10 min, 3000 g and discard the supernatant. Add 1 mL deionised water. Transfer to a new Eppendorf tube. Repeat the centrifugation and discard the supernatant. Add 1 pL of the white bacterial membrane to the target plate and dry. Add 1 pL of matrix solution and dry.	bioMerieux VITEK® MS	For Gram-positive bacteria: **88.3%** correctly identified. **85.8%** identified at genus level. **58.2%** identified at species level. For Gram-negative bacteria: **91.5%** correct identification. **92.3%** identified to genus level, **66.4%** identified to species level.	**[35]**
Centrifugation method	MALDI	For Gram-positive bacteria:	**[30]**
Inoculate 6 mL of blood culture broth into a serum separator tube. Centrifuge for 15 min, 3000 rpm Discard the supernatant and resuspend the remaining pellet above the gel separator in 3 mL sterile distilled water, mix and transfer to a new tube. Centrifuge for 10 min at 3000 rpm and discard the supernatant. Extract with ethanol and formic acid. Add 1 pL of the supernatant to the target plate and dry, then add 1 pL of the matrix solution and dry.	Biotyper	**58.7%** identified by gender. **26.0%** identified at species level. For Gram-negative bacilli; **75.8%** identified by gender. **75.0%** identified at species level. For Gram-positive bacilli; **45%** identified at genus level. For yeasts : **81.8%** identified by gender.	
Remove 4.5 mL from the blood culture bottle and dispense into 3 Eppendorf tubes (1.5 mL in each tube). Add 50 pL of Triton (10*) to each tube followed by gentle mixing. Centrifuge for 1 min, 13,000 rpm at 4°C. Resuspend each pellet from the 3 tubes with 1.5	bioMerieux VITEK® MS	**Method 1** For Gram-positive bacteria: **90%** identified at species level. For Gram-negative bacteria, **90.9%** identified at species	**[36]**

mL of physiological water. Centrifuge for 1 min at 13,000 rpm at 4°C. Remove the supernatant from the 3 tubes. Method 1: Add 50 pL of formic acid to the I^{el} tube. Method 2: add 50 pL of acetic acid to the 2eme tube. Method 3: Add 50 pL of ethanol to the 3eme tube. Add 1 pL, 2 pL or 3 pL consecutively to three different MALDI-TOF slide wells and dry fairly. Add IpL of matrix to each well and dry thoroughly.		level. **Method 2** <u>For Gram-positive bacteria</u>: **90%** identified at species level. <u>For Gram-negative bacteria,</u> **90.9%** identified at species level. **Method 3** <u>For Gram-positive bacteria</u>: **76.7%** identified at species level. <u>For Gram-negative bacteria;</u> **90.9%** identified at species level.	
Pre-raise 1.5 mL of blood culture broth. Centrifuge for 2 min, 23 g to separate erythrocytes from bacterial cells. Add 50 pL of triton X 10 to the supernatant. Mix. Centrifuge for 1 min, 15900 g. Resuspend the pellet in 1.5 mL of distilled water. Centrifuge for 1 min, 15900 g. Add 50 pL formic acid 70% and 50 pL acetonitrile. Mix well. Centrifuge for 1 min, 15900 g.	bioMerieux VITEK® MS	<u>For Gram positive bacteria</u>**:** **84%** identified at genus level. **81%** identified at species level. <u>For Gram-negative bacteria</u>**: 99%** identified at species level.	**[37]**
Place 1 pL of sumageant in triplicate on the target plate, dry and add 1 pL of matrix solution and dry.			
SELTERS Mix 1 mL of the blood culture bottle in a 2 mL mouth tube with 1 mL of 0.2% Triton X-100. Vortex for 30 sec. Centrifuge for 5 minutes at 13,000 rpm. Discard the supernatant. Add 1 mL of 0.1% DSS. Vortex to mix. Centrifuge for 5 min, 13,000 rpm. Remove the sumageant. Add 1 mL of water and vortex for 30 s. Centrifuge for 5 min, 13,000 rpm. Remove the sumageant and extract the proteins using the standard protocol: Suspend the pellets in 300 pL of water, mix and add 900 pL of ethanol. Remove the solvent after centrifugation (13,000 rpm, 2 min). Incubate the pellet residue at 37°C (for 15 min). Extract with 70% funonic acid and acetonitrile (from 20 pL to 50 pL depending on pellet size). Centrifuge and place the 1 pL supernatants on a MALDI target plate. Dry to room temperature and apply a second layer of 1 pL of ACHC matrix solution at a concentration of 10 g/ pL to the first layer and dry	MALDI Biotyper.	**87.2%** reliable identification. **13.3%** identified by gender.	**[24]**

to room temperature. Each sample was prepared in duplicate.			
Mix 1 mL of blood culture with 200 pL of 10% DSS solution. Incubate at room temperature for 5 min. Centrifuge 1 min, 13000 g Resuspend the pellet in 1 mL of HPLC water. Centrifuge for 1 min at 13,000 g. Remove the supernatant and resuspend each sample in 300 pL of HPLC water and 900 pL of 100% ethanol. Centrifuge for 2 min, at 13,000 g. Remove the supernatant. Centrifuge for 2 min at 13,000 g. Remove ethanol and dry pellets.	MALDI Biotyper	For Gram-positive bacteria: **53%** reliable identification. For Gram-negative bacteria; **89%** reliable identification.	**[38]**
for 10 min at room temperature. Performing a cthanol/fungal acid extraction Place 1 pL of each extracted supernatant in duplicate on MALDI-TOF target plates, dry and add 1 pL of ACHC matrix solution.			
Take 2 volumes of 1.5 mL from the positive blood culture vial Dispense each volume into an Eppendorf tube Centrifuge for 10 s, 600 g Aspirate 1.5 mL of the supernatant from each tube into a new Eppendorf tube. Centrifuge for 60 s, 3000 g Remove the supernatant **Step 1 :** Using a toothpick, remove the white layer of fun from the two Eppendorfs and place it on the target plate. **Step 2**: add 1.5 mL of lysis buffer to the white layer of the 2eme Eppendorf tube Mix and incubate for 3 min at room temperature. Centrifuge for 60 seconds at 3000g. Discard the supernatant Using a toothpick, remove the white layer from the Eppendorf tube and place it on the target plate.	MALDI Biotyper.	For Gram-negative aerobic bacteria **70%** identifies the level of the species after stage 1. **85%** identifies the level of the species after stage 2. For Gram-positive aerobic bacteria **41.3%,** identifies the level of the species after stage 1. **78.3%** identifies a species level after stage 2.	**[39]**
Transfer 0.75 mL of the blood culture broth into a tube containing 10 mL Tryptone Soya broth, 10 mL Triton X 100 ionic surfactant + five glass beads. Vortex for 15 sec. Centrifuge 10 min, 1300 g. Decant the supernatant with the glass beads. Replace with 10 mL sterile water. Mix and centrifuge. The glass beads trap the immiscible remains of the surfactant and help remove it when the supernatant is decanted after the first centrifugation. Decant the supernatant and invert the tube against an absorbent material to ensure removal	MALDI Biotyper	**80.25%** reliable identification. **91.1%** of enterobacteria identified. **38,5%** *Streptococcus/Enterococcus* identify.	**[40]**

of the residual water from the concentrated pellet of micro-organisms. Place 1 pL of supernatant on the target plate and allow to dry. Add			
2 pL of the matrix solution and dry.			
Remove 5 mL of the blood culture broth and transfer to a serum separation tube. Maintain the tube at a temperature of 4°C for a maximum of 4 hours. Centrifuge for 15 min at 845 g Discard the supernatant, leaving a 1 mL buffy coat. Use the buffy coat material to increase the bacterial concentration in the cell pellet. Mix 1 mL of buffy coat with 200 pL of lysis buffer. Vortex 10 s. Centrifuge 1 min, 15871 g. Remove the supernatant. Suspend the pellet with 1 mL of wash buffer. Centrifuge for 1 min at 15871g and discard the supernatant. Resuspend the pellet with 900 pL ethanol and 300 pL sterile water. Add 1 pL of the supernatant to the target plate and allow to dry. Add 1 pL of the matrix solution and allow to dry.	MALDI Biotyper	For Gram-negative bacteria; **98.4%** identifiable.	**[41]**
Centrifuge 5 mL homogenised blood culture 10 min, 700 rpm. Transfer the supernatant to another centrifuge tube Centrifuge for 15 min at 6000 rpm. Suspend the pellet with 1 mL of HPLC quality water Transfer to an Eppendorf tube and centrifuge for 15 min, 6000 rpm. Extract with ethanol/formic acid Mix the pellet with 300 pL HPLC distilled water and 900 pL ethanol. Vortex. Centrifuge for 2 minutes at 13,000 rpm. Add 25 pL of 70% formic acid and 25 pL of acetonitrile. Vortex Centrifuge 1 min, 13000 rpm. Place 1 pL of the supernatant on 4 spots of the target plate. Dry and add 1 pL of matrix solution.	MALDI Biotyper	For Gram positive bacteria: **98.4%** identifiable. For Gram-negative bacteria: **97.3%** identification liable.	**[42]**

rpm: revolutions per minute; HPLC: high performance liquid chromatography; DSS: sodium dodecyl sulphate; ACHC: alpha-cyano-4-hydroxycinnamic acid

Table IV: Steps used in sample preparation with the Rapid BACpro kit and results obtained.

Sample preparation method	MALDI-TOF equipment	Results	Reference
Rapid BACpro® II kit Mix 1 mL of blood culture broth with 500 pL of lysis	MALDI Biotyper	For Gram-positive bacteria: **100%** identified at genus	**[29]**

buffer. Centrifuge for 3 min, 3000 g. Discard the supernatant and return the pellet to 800 pL of distilled water. Add 200 pL of cationic particle solution and 200 pL of reaction buffer. Transfer the mixture to a tube and centrifuge for 1 min, 2000 g. Discard the supernatant and return the pellet with 1000 pL of 70% ethanol. Centrifuge for 1 min, 2000 g. Discard the sumageant and resuspend the pellet with 30 pL formic acid and 30 pL 100% acetonitrile. Centrifuge for 1 min, 2000 g. Add 1 pL of the supernatant to the target plate and dry, then add 1 pL of the matrix solution and dry.		level. **69.4%** identified at species level. For Gram-negative bacteria: **97.5%** identified to species level.	
Rapid BACpro® Mix 1 mL of blood culture broth with 500 pL of lysis buffer. Centrifuge for 3 min at 5000 g. Discard supernatant completely. At the time of centrifugation, 200 pL of cationic polymer suspension and 200 pL of reaction buffer are prepared in a 1.5 mL tube. The pellet is resuspended in the polymer-buffer mixture. Centrifuge for 1 min at 5000 g. Resuspend the pellet by adding 300 pL of HPLC water. Add 900 pL ethanol. Centrifuge for 2 min at 16060g. Resuspend the pellet in 20 pL -50 pL of 70% formic acid and the same amount of 100% acetonitrile. Centrifuge for 1 min, 160-60 g. Place 1 qL of supernatant on the target plate and allow to dry.	MALDI Biotyper	For Gram-positive bacteria: **82.6%** identifiable. For Gram-negative bacteria: **94%** identification liable.	**[26]**

HPLC: High Performance Liquid Chromatography

The disadvantages of these commercial kits are that they are expensive and lack sensitivity, particularly for Gram-positive bacteria. Although the purification steps, including the lysis buffers used, vary considerably between methods, the technical concepts are similar to the MALDI Sepsityper kit. Lysis buffers shown to be effective include ammonium chloride, trifluoracetic acid, 5% saponin, Triton 10X and sterile water. Several studies have compared the efficacy of these methods with commercial kits and other laboratory-developed assays, with data suggesting that using a laboratory-developed lysis and centrifugation method produces similar results to those obtained when a commercial kit is used, allowing laboratories the flexibility to choose a method that best suits their funding and staffing model **[43]**.In other studies, saponin and SDS have been shown to perform similarly to or better than the Sepsityper kit**[23]**.The SELTERS method, which offers a single extraction step, significantly reduced sample handling time and gave comparable or better results than the Sepsityper kit, with a sample cost that was approximately 10 times lower than the cost of the commercial kit and identical consumption of consumables **[24]**.

When assessing the effectiveness of MALDI-TOF on Gram-positive, Gram-negative bacteria and mycetes, there is a slight difference: identification is better for Gram-negative bacteria than for Gram-positive bacteria. As observed by other studies and noted in the MALDI-TOF limitations reported by the manufacturer, the lower yield of valid MALDI-TOF results with streptococci and staphylococci may be due to the difficulty in distinguishing between closely related group members of streptococci belonging to the *S. mitis* group (i.e. *S. pneumoniae, S. mitis*, *S. sanguinis*, *S. oralis*), which did not produce very high scores. In addition, the composition of the cell wall of Gram-positive bacteria conferring increased resistance to lysis and the possible presence of certain residual blood proteins could be considered as a limit to correct identification by MALDI-TOF. In fact, some authors have proposed ultrasound treatment for better disruption of the peptidoglycan layer of the cell walls of Gram-positive bacteria**[44]**.

MALDI-TOF can provide identification results in less than an hour using purified bacterial clots obtained from blood culture flasks, with mean times before and after introduction of MALDI-TOF of 33 h and 7 h respectively**[45]**.

With the Sepsityper kit, Gram-negative bacilli had a higher identification rate than Gram-positive cocci. The number of unacceptable samples was high for *Streptococcus* species. *S. pneumoniae*, *Bacteroides fragilis* and yeasts were also the most difficult to identify. The overall percentage agreement of this kit at species level for Gram-positive species ranged from 33% to 56.8% and for Gram-negative species from 51% to 85.2%. For anaerobes, the Sepsityper method gave the best results for Gram-positive bacteria. However, the saponin method gave better results in identifying these germs. This kit does not identify all polymicrobial samples.

In addition to a 10-minute shorter run time per sample than the Sepsityper kit, the Rapid BACpro kit was able to identify a greater number of Gram-positive bacteria. However, there were no significant differences in the identification of Gram-negative bacteria. Rapid BACpro allows more accurate identification of bacteria at species level than the Sepsityper kit. Both kits have difficulties in identifying groups of *Staphylococcus* and *Streptococcus*.

For the internal methods, saponin gave results that were superior to the centrifugation method and comparable to the Sepsityper identification method.

A second type of method developed in the laboratory involves inoculating positive blood culture bottles with blood gel medium and incubating at 35°C for a short period (3-5 h) to promote microcolony growth, followed by MALDI-TOF analysis. Although this method does not provide results as quickly as commercially available extraction kits, the growth of microcolonies can be considered as normal growth and used for standard MALDI-TOF identification. This process reduces the associated work that would be required for extraction kits. However, as the short incubation time does not allow colony morphology to develop fully, polymicrobial cultures cannot be detected by this method **[43]**. In this study, Fang et al**[46]** compared two rapid protocols, direct identification from blood culture bottles (Sepsityper kit) and identification after 5-6h early subculture by MALDI-TOF of microorganisms in paediatric patients. For the

protocol using the Sepsityper kit, 85.2% and 64.3% of the micro-organisms were correctly identified at genus and species level. For the early subculture protocol, 89.6% and 70.4% of the microorganisms were correctly identified at the genus and species level, respectively. In addition, the early subculture protocol showed similar performance between Gram-positive and Gram-negative bacteria (at the genus level: 93.7% (Gram-positive) versus 87.9% (Gram-negative); at the species level: 68.4% (Gram-positive) versus 81.8% (Gram-negative)). In addition, the Sepsityper kit protocol showed similar performance between Gram-positive and Gram-negative bacteria (at the genus level 86.1% (Gram-positive) versus 84.8% (Gram-negative)). However, the Sepsityper kit protocol showed better performance of Gram negatives at the species level (58.2% (Gram positive) versus 81.8% (Gram negative). Both rapid protocols showed better performance at the genus and species level when the blood culture positive time was less than 19 h. There was no significant difference between the Sepsityper kit protocol and the short-term subculture protocol.The Sepsityper kit protocol does not require 5-6 hours of subculture and is faster than the short-term subculture protocol, requiring less than 60 minutes for a rapid result from the time a positive blood culture is detected.The early subculture protocol does not require additional costs for the purchase of the commercial kit and additional sample preparation, such as washing, centrifugation and extraction steps. Its ease of use means that it can be used in all clinical laboratories equipped with MALDI-TOF[46].
Tsai et al**[47]** attempted in this study to develop a method with a considerably shortened turnaround time for clinical application by identifying the optimal incubation period of a subculture made from a positive blood culture bottle. The optimum incubation time for confirming bacterial identification of Gram-negative organisms was identified as 3.5h. *Enterococcus* spp. were 100% identified after 2.5h of culture, which is the shortest incubation time to identification among Gram-positive bacteria; only 68.2% of *Staphylococcus* spp. were detected after 2.5h. The optimum incubation time for Gram-positive bacteria was 4.5h. For antibiotic susceptibility testing, studies are constantly focusing on improving rapid susceptibility testing. In this study, the antibiogram is determined by the Phoenix M50 system after identification by MALDI-TOF (Bruker Daltonik, Bremen, Germany). This approach has significantly reduced the average turnaround time from 48h to 28h for reporting bacterial identity and has reduced the time for susceptibility testing from 72h to 50.3h compared to conventional methods. As a result, this approach allows a physician to prescribe the appropriate antibiotic(s) ~21.7h earlier, improving patient outcomes **[47]**.
Fang et al. reported the use of MALDI-TOF to identify positive haemocultures from a short incubation of a sheep blood gelose subculture pre-warmed to 5%, then incubated for 2, 3 and 4 h at 35 C. Results were compared to MALDI-TOF identification from routine cultures incubated for 18 to 24 h. Identification was concordant for 73.8% of subcultures. The monomicrobial cultures with Gram-negative bacteria were mainly enterobacteria, and concordant identification within 4 h of incubation was obtained for 88.9% of Gram-negative vials. For Gram-positive bacteria, a concordant identification

was obtained in 62.1% of Gram-positive flasks, and 85.7% of *S. aureus* flasks were concordant compared with 50% for SCN. In this study, they found a rate of discordant identifications of 2.4% for monomicrobial vials and 10% for polymicrobial vials. Early identification from early subcultures is not suitable for anaerobic or slow-growing micro-organisms, such as yeasts. When these isolates are indicated by microscopy, alternative methods such as MALDI TOF directly from positive vials may be required **[48]**.

Zabbe et al**[49]** evaluated this method on colonies obtained on chocolate gelose after 3h incubation. A portion of the bacterial growth veil was deposited in duplicate on the MALDI-TOF target plate and processed using Biotyper 3.0 software. The vast majority of enterobacteria (92%) were correctly identified at species level. For staphylococci, 58% were correctly identified at species level, while most *Pseudomonas* isolates were correctly identified. Conversely, anaerobic bacteria, yeasts and fungi were poorly identified by MALDI-TOF. Taking all species together, MALDI-TOF was able to identify 77% of isolated bacteria.This early result led to a real change in treatment regimes in 20% of the patients tested.The performance was excellent for enterobacteria and enterococci.The results were less good for anaerobic bacteria due to the atmosphere chosen for this rapid incubation.In addition, the results were not very good for yeasts either. The success rate is similar to the results obtained in various studies using extraction methods with commercial kits applied directly to the positive blood vials without a 3-hour incubation period. However, unlike this method, these kits are expensive and the method is time-consuming, requiring a dedicated full-time technician to carry out the various steps, including lysis and centrifugation.Other studies have reported similar percentages of correct identification with MALDI-TOF on 5-hour cultures from positive blood vials **[49]**.

Halavaara et al**[50]** evaluated the impact of MALDI-TOF after a short incubation period of 3-4h on bacteria caused by *P. aeruginosa*, *Enterococcus* spp. and Amp-C-producing Enterobacteriaceae. Identification by MALDI-TOF resulted in a 12.9% increase in cases receiving appropriate antibiotic treatment within 48 hours of blood culture collection.The importance of rapid identification was highlighted in bacteremia caused by enterococci, where MALDITOF led to appropriate antibiotic treatment in 87.7% of cases.Of all cases, antibiotic treatment was appropriate in only 23.6% of cases within 12 hours of blood culture collection, but improved to 72.6% within 48 hours. In general, the median time between blood culture collection and receipt of appropriate antibiotic treatment was 33.5 hours**[50]**.

Curtoni et al. evaluated an easy and rapid protocol for direct MALDI-TOF MS identification of micro-organisms from a positive blood culture on solid medium at different incubation times (3h, 5h, 24h). MALDITOF concordance with conventional methods at species level was 60.5%, 80.2% and 93.8% at 3, 5 and 24h, respectively. Identification performance at the species level was 64.1%, 85.0% and 94.1% at 3.5 and 24 h, respectively. For cocciGram positives, *S. aureus* had the best identification results, while the worst results were obtained with the streptococci group. They observed good results for SCN after only 5 h of incubation with no misidentification and an identification rate at 5 h of 88.2% at species level. For Gram-negative bacteria, enterobacteria had the highest identification rate, while unsatisfactory results were obtained for non-fermenting Gram-negative bacteria as well as for Gram-positive bacilli and yeasts with no identification at species level even after 5 h of incubation. However, for Gram-positive bacilli, this must be considered in terms of clinical relevance, as they are often considered contaminants. These unsatisfactory results are probably due to the low intrinsic growth rate of these micro-organisms**[51]**.

Kock et al**[52]** compared the applicability of MALDI-TOF-based species identification from short-term incubated cultures in routine laboratory practice with conventional diagnostics and evaluated its effect on empirical antibiotic therapy.Results: The median time from detection of vials as 'positive' by incubators to subsequent processing was 6h 21 min. The median time between microscopy and communication of results to the department was 15 minutes.species results were available to clinicians after a median of 188min (231min for Gram-positive bacteria, 151min for Gram-negative bacteria) compared with 909min when conventional diagnostics were used.The majority of vials were flagged as 'positive' by the automated incubator outside laboratory operating hours, and the median time to further processing of a vial (i.e. microscopy and plating onto solid support) was >6h. In addition to offering longer service hours, another option to address this issue could be the installation of blood culture bottle incubators in areas of the hospital available to clinical staff so that bottles could be continuously loaded into the systems.They also found that clinicians changed empirical antibiotic therapies in 20% of cases after reporting the results of the species test. This was more

On the other hand, a majority of antibiotic therapies remained unchanged This could be explained, firstly, by compliance with well established guidelines for empirical therapies covering the species reported. Secondly, by the difficulties experienced by the treating physicians in correctly interpreting the species report (without antibiotic susceptibility testing) in terms of how to adjust empirical therapies on the pretext that they are more precise and sophisticated, or thirdly, without taking into account the diagnostic results**[52]**.

Rapid identification of bacteria using MALDI-TOF has already been incorporated as a routine test in many institutions; however, no standardised clinical pretreatment method is available, so each institution seems to use its own methods **[43]**.

1.1.1.1. Sensitivity tests using MALDI-TOF

Sauget et al. have developed a technique for the rapid detection of *Escherichia coli* resistant to amoxicillin and cefotaxime (CTX) from MALDI-TOF-based positive blood cultures. The MALDI Biotyper-Antibiotic *Susceptibility Test Rapid Assay* method indirectly estimates the growth of bacterial strains. It involves incubating the isolate in the presence and absence of an antibiotic. If the antibiotic is active, the intensity and number of peaks detected will be reduced. A growth rate is then calculated from the spectra obtained **[53]**.Anantharajah et al. also used this method to identify Gram negative bacteria producing carbapenemases **[54]**. Axelsson et al. used this method to test the sensitivity of *E coli* and *Klebsiella pneumoniae* strains producing extended-spectrum debeta-lactamases (ESBL) to CTX, ciprofloxacin and meropenem**[55]**.

The test can accurately determine the sensitivity of *E. coli* to amoxicillin and identify CTX-susceptible *E. coli* directly from a positive blood culture in less than 4 hours. However, it cannot be used to test for piperacillin-tazobactam sensitivity in the absence of the necessary data on optimal test conditions. The development of automated commercial kits with the most important antibiotics would be useful for rapidly orienting the treatment of high-risk patients **[53]**. An incubation time of 90 min can detect antibiotic resistance in *E coli* and *K. pneumoniae*, but for Gram-positive bacteria data have shown that the time should be extended by at least 180 to 300 min depending on the concentrations of antibiotic selected **[55]**.

To identify carabapenemase-producing enterobacteria, Yu et al **[56]** also studied the effectiveness of MALDI-TOF in detecting their hydrolysis activity. To do this, a bacterial suspension was prepared in the presence of ertapenemes (ertapenemes, meropenemes and imipenemes are used as carbapenemes), and injected into negative blood culture bottles. After incubation in Bactec, the flasks are then analysed by MALDI-TOF. Carbapenemase producers can be distinguished by the absence of intact ertapenem peaks. The presence of peaks corresponding to the hydrolysed and decarboxylated form of ertapeneme at high intensity. This is a fast (results are provided within 3 hours), accurate, efficient and cost-effective method with high specificity and sensitivity. However, this method cannot be used to detect all carbapenemases. Certain isolates of enterobacteria positive for the NDM-1 gene have not been distinguished as carbapenemase producers. Furthermore, if 2 species of enterobacteria are present in the same vial, one of which is resistant, all the strains will be identified as carbapenemase producers **[56]**.

Nix et al. described a new method for testing antibiotic susceptibility based on the *direct-on-target* microdroplet *growth assay* (MALDI-TOF) for Gram-positive bacteria directly from positive blood culture broth using methicillin resistance in *Staphylococcus aureus* as an example. In this study, they compared the different processing steps of blood culture bottles: dilution, lysis/centrifugation and differential centrifugation, as well as the performance of each technique after incubation of the target plate for durations of 3, 4, 5 and 6h.

After 4h incubation, the lysis/centrifugation technique gave the best result with 100% precision of resistance detection and 96.4% validity. These same results were obtained after 6h incubation with the dilution and lysis/centrifugation techniques **[57]**.

Idelevich et al. also used this method to study carbapenem resistance in enterobacteria and concluded that the best results were obtained with the lysis/centrifugation method after a 4h incubation with a validity of 96.3% and a sensitivity of 91.7%. Centrifugation methods require the attention of the technician and are difficult to automate, whereas dilution methods can be performed with a minimum of handling time. In addition, it is important to prepare a clean sample and standardised inoculum before starting the test **[58]**.

2.3.2. Techniques based on chain polymerisation

Polymerase chain reaction (PCR) amplification is based on an *in vitro* enzymatic reaction that enables genomic deoxyribonucleic acid (DNA) or complementary DNA (cDNA, generated by reverse transcription from a molecule of ribonucleic acid (RNA)) sequences to be amplified several million times.Jungho et al. described a simple and easy method for extracting DNA by paper chromatography that requires just two reagents and takes just 10 minutes.

The Punch-it™ NA-Sample kit could be offered as a rapid method.This assay is designed to easily isolate nucleic acid. It uses the paper chromatographic method to effectively remove PCR inhibitors from samples. Purified nucleic acids bound to the membrane can be used directly in PCR or real-time PCR reactions as templates. This method is accurate and convenient for extracting bacterial and fungal DNA from blood culture bottles, and the sensitivity and specificity of the kit were 95.4% and 100% respectively. A comparison of extraction by the kit and extraction by the boiling method showed that the Punch-it™ NA-Samp lea kit led to more accurate results for the identification of 3 bacteria *S. aureus*, *Bacillus spp*. and *E.* faeciumbut one sample of *P. aeruginosa* was not detected by this method [59].

2.3.2.1. MagicPlex

Currently, PCR tests dedicated to bacterial diagnostics are mainly multiplexed and performed in automated systems to ensure a smoother laboratory workflow. The Magicplex™ system can detect 90 pathogens and 3 resistance markers within 6-7 hours. To achieve final identification, a three-step procedure is required, as an initial screening PCR amplification is performed, followed by two real-time PCR steps.The Magicplex™ has increased the number of cases diagnosed in adults and children by 110% and 143% respectively **[60]**.

2.3.2.2. LightCycler® Septifast

The LightCycler® Septifast (Roche Molecular System, Basel, Switzerland) has been the subject of numerous studies. This molecular test, which allows the detection of 25 pathogenic agents including 6 fungi in approximately 6 hours, has revealed a high false positive rate and it appears that the performance of Septifast varies according to the population studied. False positives are higher in immunocompromised patients with pneumonia or septicaemia**[60]**.

2.3.2.3. Multiplex tandem PCR

Briony et al. evaluated the performance of 12 multiplex tandem PCRs containing targets for the identification of staphylococci, streptococci and enterococci. Identification by this technique was consistent with phenotypic identification at the genus level in 96.6% and at the species level in 100% of single microbial samples **[61]**.

Methicillin and vancomycin susceptibility results determined by multiplex tandem PCR in blood culture broths showed complete concordance with those determined by phenotypic methods for *S. aureus* and *E. faecium* isolates.

Gram-positive bacteria and their major antibiotic resistance markers were reliably identified with the multiplex tandem PCR assay within 3 hours of a positive blood culture result. The presence or absence of methicillin resistance was correctly determined in 100% of *S. aureus* isolates, and the presence or absence of vancomycin resistance mediated by the *vanB* gene was correctly detected in 100% of *E. faecium* isolates.

This test was highly accurate in detecting methicillin resistance caused by *the mecA* gene. Variants such as emerging mecC-positive methicillin-resistant *Staphylococcus aureus* (MRSA) would not be detected due to sequence differences. However, mecC-positive MRSA is currently rare and its detection is a problem that would affect most, if not all, commercial and in-house molecular tests designed to detect *mecA-derived* methicillin resistance**[61]**.

2.3.2.4. Multiplex PCR

Rocchetti et al also evaluated a standardised multiplex PCR to detect Staphylococcusspp, *S. aureus* and methicillin resistance encoded by the mecA gene directly from blood culture bottles. In this study, 23% were positive *for* S. *aureus* and 77% were positive for SCN, 50.6% of *S. aureus* strains and 82.5% of SCN strains were resistant to methicillin. 45.1% of *S. aureus* isolates and 72.5% of SCN isolates were resistant. Other studies have also reported high levels of methicillin-resistant *Staphylococcus* spp. Given this increase in resistant strains, treatment options have been reduced to glycopeptides and new synthetic drugs such as linezolid, which not only are expensive and toxic, but also select for resistant strains.

The multiplex PCR test proved to be sensitive, specific, rapid and less expensive than the phenotypic technique used, and showed good agreement with the phenotypic results**[62]**.

2.3.2.5. PCR-dipstick

Rathina et al. evaluated the performance of PCR-dipstick to detect vancomycin-resistant MRSA and *Enterococcus*. The PCR dipstick is a multiplex PCR-based detection technique where DNA-DNA hybridisation is used and results are interpreted in Trail nu.It was designed to target three antibiotic resistance genes: *mecA* in MRSA and *vanA/vanB* in vancomycin-resistant *Enterococcus* from positive blood culture samples. The PCR strip showed 100% sensitivity and specificity in detecting the *mecA*, *vanA* and *vanB* genes in approximately 3h. The sensitivity of this technique was 10 times greater than that of agarose gel electrophoresis in the detection of PCR amplicons. The PCR strip could be used to diagnose vancomycin-resistant MRSA and *Enterococcus* from positive blood culture samples at a low cost of around $4 per reaction, which is relatively low compared to other detection techniques such as real-time PCR. Other blood culture-based molecular detection systems, such as real-time PCR and DNA microarrays, require a sophisticated infrastructure and a huge amount of equipment to implement in routine clinical examinations, particularly in low-resource healthcare facilities. Interpretation of results is very easy for dipstick PCR and involves visualization by the naked rail as shown in **Figure 3** without the need for additional equipment, such as a computer with special software.This technique is very simple and easy to perform.The limitation of dipstick PCR for vancomycin-resistant MRSA and *Enterococcus* is that it requires isolation of DNA from positive blood cultures which is then subjected to dipstick PCR [63].

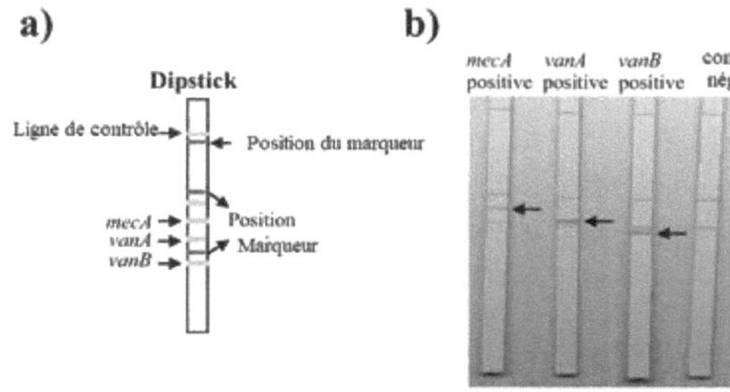

Figure 3: Visual interpretation of resistance genes by PCR-dipstick[63] a) position of the probe for the 3 different resistance genes *mecA*, *vanA* and *vanB* printed on the strip.b) identification of the 3 resistance genes in the different positions on the strip.

2.3.2.6. FilmArray®

The FilmArray identification system® (BioFire Diagnostics, Salt Lake City, USA)*blood culture identification* (FA-BCID) is a fully automated PCR device, from extraction to data analysis. Amplification is based on multiplex PCR and DNA is analysed using melting curves. The identification panel includes 24 microorganisms, including 5 *Candida* species and 3 resistance genes (*mecA*, *vanA/vanB* and *blaKPC*).

BCID2, a new update of the original BCID, identifies 33 species, including a new distinction between *E. faecalis* and *E. faecium*. In addition, it analyses 10 genetic resistance markers, including the common genes encoding carbapenemases and the genetic marker for colistin resistance (*mcr-1*). Initially designed to run one sample per cycle, the FilmArray® Torch has recently been developed and can accommodate up to 12 modules. The main advantage of the FA-BCID is the minimal handling time (5 min), while the time to result is only one hour.the sensitivity for detecting the micro-organisms included in the panel is less than 96% with a specificity of 99% **[60,64]**.

Ray et al. studied the effectiveness of this method in the management of septic children. FA-BCID results were judged to modify clinical management in 63 of 117 episodes (54%). Antibiotics were started/modified in 23 episodes (19%) and decreased/discontinued/stopped in 29 episodes (25%). A percentage of 10% of children were discharged from hospital early, saving a cumulative total of 14 bed days**[65]**.

Salimnia et al. presented the results of a trial carried out in eight centres comparing the panel's sensitivity and specificity with those of standard phenotypic identification techniques, as well as with the molecular methods used to distinguish

Acinetobacterbaumannii from other bacteria in the *Acinetobacter* genus and to detect antibiotic resistance genes.The sensitivity of blood culture identification for almost all the bacteria detected exceeded 96%; the sensitivity and specificity for *vanA/B* and *blaKPC* were 100%; those for *mecA* were 98.4% and 98.3%, respectively.FA-BCID is capable of detecting microorganisms in mixed cultures.however, other studies have shown that the presence of multiple microorganisms favours discordant results.There are several possible sources of false-positive results in nucleic acid-based tests, as molecular methods can detect genetic material present in non-viable organisms.FA-BCID results that do not agree with the Gram stain result of the vial should be evaluated carefully **[66]**.

Verhoeven et al. evaluated the performance of the FA-BCID multiplex test against a protocol based on rapid MALDI-TOF identification after 4 hours of subculture. Overall, 137 (91.3%) pathogens were correctly identified by FA-BCID at species and genus level compared with 117 (78.8%) by MALDI-TOF identification, confirming the excellent sensitivity of the FA-BCID test, particularly in cases of multi-microbial infection.Four out of 22 *S. aureus* isolates (18.2%) and 35 out of 41 SCN isolates (85.4%) were found to be positive for *genemecA* by FA-BCID.91.5% overall concordance of FA-BCID with the reference technique was achieved. This result was significantly better than that obtained with the rapid MALDI-TOF test (79.7%). The sensitivity of the FA-BCID test was 90.9%. This is higher than that of the rapid MALDI-TOF method (87.2%).

Due to the limited number of targets included in the test, it must be coupled with another identification strategy. Its high cost and the need to test each sample individually may represent an additional limitation in high throughput laboratories. Although less sensitive, particularly for slow-growing bacteria, the fast MALDI-TOF method used as a comparator in this study represents a much cheaper alternative.As suggested by Fiori et al, 9, it may be worth reserving the FA-BCID approach for those vials for which identification by the rapid MALDI-TOF method has failed (-20% of vials tested in this study), and reserving this identification approach for patients considered to be the most critical**[67]**.

Payne et al. also evaluated FA-BCID against MALDI-TOF and standardised identification with antibiotic susceptibility testing. FA-BCID gave accurate identification in 91% (102/112) of cases. MALDI-TOF gave accurate identification in 81% (91/112) of cases. The average time to obtain an identification result was 2.4h and 2.9h for FA-BCID and MALDI-TOF respectively.44 FA-BCID and MALDI-TOF results were followed by an antibiotic change based on rapid identification and detection of resistant pathogen genes.To provide rapid identification and detection of resistance genes in the most cost-effective way, MALDI-TOF can be used initially, with the FA-BCID test reserved for certain indications, such as suspected *S. aureus*, streptococci or *Enterococcus* spp.bacteremia, candidemia and polybial cultures. The cost-effectiveness of testing for suspected multiresistant Gram-negative bacteria (blaKPC) will depend on the local epidemiology of each laboratory **[68]**.

Verroken et al. evaluated the impact of rapid identification by FA-BCID on the management of critically ill patients. 96.2% of bacteria were identified by FA-BCID, covering 85.2% of the epidemiology of ICU strains. The median identification time for positive blood culture bottles using the FA-BCID test was reduced to 1h35 compared with the median time for MALDI-TOF. The molecular test detected the *mecA* gene with 100% sensitivity and specificity. Similarly, the 7 enterococcal strains were vanA/Bnegative and the 59 Gram-negative strains were blaKPCnegative and were in concordance with the routine antibiotic susceptibility test, with no false-positive results observed.The FA-BCID test led to a substantial reduction of 10h in the time taken to administer an antibiotic treatment. The results of the FA-BCID improved the therapeutic management of 31 out of 110 patients and resulted in a 16h reduction in the time to initiation of active treatment compared with routine tests **[69]**.

Fhooblall et al. evaluated the effectiveness of the performance characteristics of the FA-BCID kit compared with the conventional technique used for the identification and antibiotic susceptibility of bacteria isolated from blood cultures. The detection of bacteria by FA-BCID was 92.6% accurate, and the detection of antibiotic resistance markers was 100% consistent with the reference method. Blood cultures with a single organism were identified with 93.8% accuracy, while blood cultures with more than one organism were identified with lower accuracy (85.7%)**[70]**.

Pardo et al. evaluated the impact of the FA-BCID panel on the management of patients with Gram-positive cocci and *Candida* blood cultures. Clinical and economic results were evaluated retrospectively. FA-BCID identification of contaminating SCNs reduced length of stay and saved approximately $30,000 per 100 patients tested. FA-BCID led to a shorter duration of empiric vancomycin therapy for patients with contaminated blood cultures and methicillin-susceptible aS. *aureus* bacteremia. Patients with vancomycin-resistant *Enterrococcus* bacteremia received active treatment earlier than controls. FA- BCID, combined with pharmaceutical intervention, was a cost-effective tool for improving patient care**[71]**.

Seito et al. evaluated the FA-BCID panel using 54 positive blood cultures, *mecA* genes were detected in 12 species of *Staphylococcus*, all of which showed resistance to methicillin in susceptibility testing, while blaKPC, *vanA* and *vanB* genes were not detected, in agreement with the results of antibiotic susceptibility testing **[72]**.

2.3.2.7. GenMark Dx ePlex

The GenMark Dx ePlex blood culture identification panels are fully automated PCR-based tests designed to identify Gram-positive and Gram-negative bacteria, fungi and bacterial resistance genes within 1.5 hours from a positive blood culture. The choice of panel(s) (Gram-positive, Gram-negative and/or fungal pathogens) is determined by the result of microscopic observation after Gram staining of the positive vials. ePlex can identify 20 genera and species of Gram-positive bacteria and 4 antibiotic resistance genes (BCID- GP panel), 21 genera and species of Gram-negative bacteria and 6 antibiotic resistance genes (BCID-GN panel), and 16 fungal genera (BCID-FP panel) from a positive blood culture. Huang et al. evaluated the performance of this method compared with the routine method.the sensitivity for detection by the ePlex panels of target culture isolates was 97% and 99% for the Gram-positive panel and 99% and 96% for the Gram-negative panel.samples with target resistance results were correctly detected by the ePlex panels. The results were very accurate and rapid, which optimises the clinical management of septic patients. ePlex is an excellent diagnostic tool for the rapid identification of pathogens responsible for bloodstream infections **[73]. Figure 4 shows** the steps involved in performing the ePlex test.

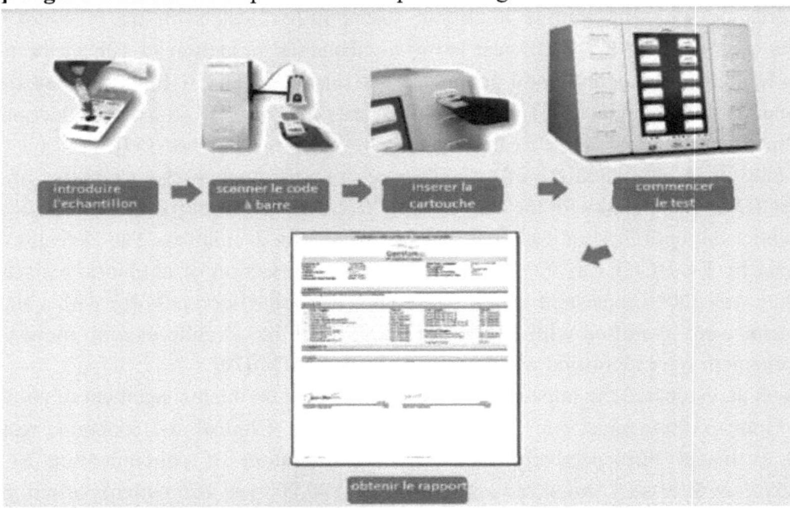

Figure 4: ePlex test execution process [from 74].

Carroll et al. evaluated the clinical performance of the BCID-GP panel against conventional methods for the identification and detection of antibiotic resistance genes for common Gram-positive bacteria recovered from positive blood. The overall sample accuracy for the panel organisms was 89% before resolution of discordant results. The ePlex BCID-GP panel compares favourably with targeted molecular methods for the identification of 20 Gram-positive bacteria and four antibiotic resistance genes in positive blood culture bottles. It should be noted that if there are several Gram-positive cocci in a sample that is also positive for a resistance marker, there is no way to link the gene to the species detected, which would probably occur most often with *mecA* when the sample contains *S. aureus and one or more other* Gram-positive organisms. The panel is very inclusive, only 2.2% of clinical isolates identified by conventional methods were not targeted by the BCID-GP panel, in contrast to the 7.5% to 11.9% reported for other currently available commercial tests, most of which are opportunistic organisms that are sometimes associated with significant bacteremia.In addition, identifications at genus level as well as further identification at species level may offer the potential for more targeted therapeutic interventions based on the antibiogram, for example, the BCID-GP panel may provide a result on the

Enterococcus genus, which may prompt the laboratory to conduct further investigation to exclude less common *Enterococcus* species.*Clostridium* species were among the Gram-positive bacteria not targeted by this panel and are generally the second most common anaerobic bacteria isolated from blood cultures. Detection of resistance genes and/or rapid phenotypic tests are important for optimising treatment. Several studies have shown that when either of these elements is combined with interventions, the median time to optimal therapy, particularly for methicillin-susceptible *S aureus* and *Enterococcus* species, can be shortened from 14 h to more than 24 h. In this study, the prevalence *of vanA* was low among prospective samples, and no vanB-containing enterococci were detected. Distinguishing *vanA* from *vanB* is important for predicting glycopeptide susceptibility, as *vanA-containing* enterococci are resistant to both vancomycin and teicoplanin, while *vanB-containing* enterococci remain susceptible to teicoplanin. ThePlex system is the first commercially available rapid multiplex PCR panel capable of identifying *Staphylococcus lugdunensis* as well as the presence of a *mecA* gene.*S. lugdunensis* is an important staphylococcal species to identify due to its more aggressive nature than other SCN.However, the BCID-GP panel detected all organisms identified by comparative methods in 73% of polymicrobial samples.The GenMark Dx ePlex system uses three panels for the identification of Gram-positive, Gram-negative and fungal organisms, all of which are now commercially available. The use of multiple panels allows the inclusion of a wider range of targets, but the selection of the appropriate panel(s) depends on the precise results of the Gram stain**[75]**.

Wolk et al. report the results of a multicentre clinical study evaluating the performance of the ePlex BCID-GN Gram-negative panel. One of the most important advantages of BCID-GN is its ability to differentiate between bacteria with similar Gram reactions, thereby enhancing its ability to identify coinfections. BCID-GN contains targets for class A carbapenemase, KPC, class B metallo-beta-lactamases (IMP, VIM and NDM) and class D beta-lactamase OXA.BCID-GN targets the CTX-M group of ESBLs, which are the most common ESBLs worldwide and induce resistance to penicillins, cephalosporins and monobactams. The accuracy of the antibiotic resistance target genes was high, ranging from 93.1% to 100%, and had a corresponding performance similar to or better than that of phenotypic and molecular comparison methods for antibiotic susceptibility testing.A limitation of BCID-GN is the non-inclusion of the *mcr-1* and *mcr-2* genes, which are indicative of colistin resistance.BCID-GN offers advantages as a significant aid in the diagnosis of specific bacterial agents. BCID ePlex panels are becoming the most comprehensive rapid molecular diagnostic offering for the identification of bacterial pathogens and resistance mechanisms, with the ability to detect a total of 41 bacterial targets, 15 fungal targets and 10 antibiotic resistance markers from a positive blood culture bottle **[76]**.

2.3.2.8. VERIGENE®

The VERIGENE® system uses Nano Grid technology. Unique gold nanoparticle probes designed to detect target nucleic acids and proteins. The gold nanoparticles used in the *Nano Grid Technology* are functionalised with a number of antibodies specific to the target protein or a defined number of oligonucleotides.The technology used in the VERIGENE® system enables automated nucleic acid extraction, PCR amplification and transfer of eluted nucleic acids for primary and secondary hybridisation. The instruments used in the VERIGENE® system generate automated qualitative analysis of the results**[77]**. The VERIGENE® system requires less than 5 minutes of hands-on time for all tests. **Figure 5** describes the stages of analysis using the VERIGENE® technology **[78]**.

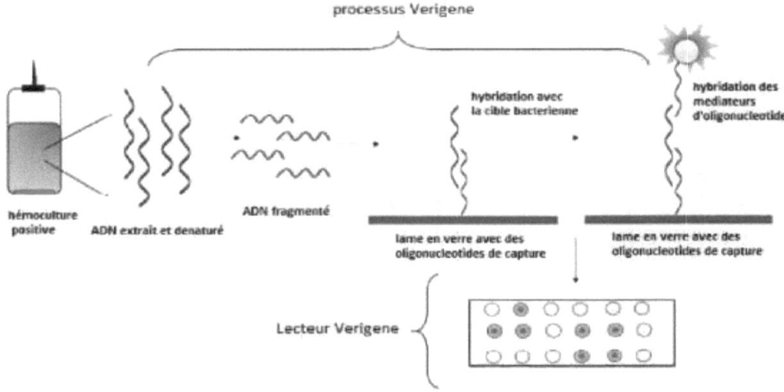

Figure 5: Stages of analysis using VERIGENE® technology [78].

VERIGENE® is a nucleic acid microarray that can be used to provide rapid information on positive blood cultures. It detects genetic information that can reveal the genus/species of bacteria and the presence of various resistance genes. Different matrices are used to study Gram-positive or Gram-negative blood cultures. The test has a turnaround time of around 2.5 hours, allows rapid identification of the pathogen and its antibiotic resistance markers simultaneously from positive blood cultures and works in parallel with traditional methods. The initial acquisition cost of the VERIGENE® analyser is considerable (~$20,000). Subsequently, each test costs around $50 to $100. Given the possibility of avoiding or restricting antibiotic therapy, this tends to reduce costs. Secondary benefits include limiting exposure to antibiotics, thereby reducing antibiotic resistance and consequently *C. difficile* infection. VERIGENE® identifies pathogens with high sensitivity (~97%) and good specificity (~99.5%)**[79]**.

Guevara et al. evaluated the VERIGENE® GramPositive (BC-GP) and Gram-Negative (BC-GN) *Blood Culture* nucleic acid assays for rapid and accurate detection of pathogens compared to conventional methods.The BC-GP assay consists of 12 bacterial targets (9 different bacteria are identified at species level, 3 are identified at genus level) and three resistance markers. It can detect genera such as *Staphylococcus, Streptococcus, Enterococcus* and *Listeria* as well as specific resistance markers such as *mecA*, *vanA* and *vanB*.The BC-GN test was recently approved by the *Food and Drug Administration* (FDA) to have 8 bacterial targets (4 different species and 4 different genera) and six resistance markers, including CTX-M and OXA. It originally contained 9 and six resistance markers. The accuracy of VERIGENE® was 98.7% and 98.4% for Gram-positive and Gram-negative cultures, respectively. The VERIGENE® nucleic acid test has made a significant contribution to reducing the time required to detect the causative pathogens, as well as initiating appropriate treatment earlier.An important criterion for rapid blood culture tests such as VERIGENE® is the ability to differentiate between different isolates within a given polymicrobial culture, but the accuracy of identification remains reduced compared to monomicrobial samples. This study shows that VERIGENE® increased the detection and identification of Gram-positive bacteria by 1.3% compared to VITEK® **[77]**.

The limitations of this study, including limitations noted in the Luminex VERIGENE® package insert, include that the test is not approved for testing outside of FDA-approved targets in the USA and false positive results due to cross-reactivity of and *S. pneumoniae* that have homologous genetic material.For almost all isolates, Luminex VERIGENE® BC-GP was as informative as traditional culture identification, and resistance markers were identified with a very high rate of inaccuracy, with only 1% of results discordant.Providers can therefore use this technique with confidence.For the rare cases where there was disagreement, the clinical impact was minimal and did not expose patients to an increased risk of harm**[80]**.

Hayakawa et al. studied the impact of rapid identification of positive blood cultures using the VERIGENE® system on antibiotic prescribing, with changes in antibiotic prescribing occurring in 27.6% of patients after VERIGENE analysis® and 37.8% of patients after conventional identification and sensitivity testing. Prescription changes after VERIGENE® identification were more frequent for Gram-positive bacteria (41.5%) than for Gram-negative bacteria (23.5%). Among bacteria caused by a single pathogen targeted by the VERIGENE® test, the concordance of bacterial identification between the two tests was 97.4% for Gram-positive bacteria and 100% for Gram-negative bacteria. The times between the start of incubation and reporting of results for the VERIGENE® system and the conventional test were 28.3 h and 90.6 h respectively. In only four of the seven community-acquired bacterial outbreaks in which two isolates were identified by conventional tests did the VERIGENE® test correctly identify both organisms. The test correctly predicted target antibiotic resistance.It was more difficult to narrow down the antibiotic regimen for Gram-negative organisms because VERIGENE® BC-GN was unable to predict resistance patterns other than CTX-M ESBLs and carbapenemase-producing bacteria, such as resistance to quinolones and/or aminoglycosides, or detailed phenotypes for betalactamins.The results of this study indicate that caution should be exercised when implementing VERIGENE® tests. Firstly, VERIGENE® relies on accurate Gram staining results to configure the correct panel. In this study, an outbreak of community-acquired *S. pneumoniae* was falsely identified as community-acquired due to the use of a Gram-negative panel, resulting in missed detection by VERIGENE® BC-GP.In addition, the performance of the VERIGENE® test for the correct identification of multiple target organisms isolated from the same blood culture flask was sub-optimal and this was attributed to an insufficient quantity of target bacteria in the cultures**[81]**.

Freiberg et al. evaluated the identifications and antibiograms of isolates that were not identified by theVERIGENE® BC-GN or BC-GP test.This study provides important information on the strengths and limitations of this method.Previous studies have shown that approximately 5-8% of Gram-positive blood culture isolates and 5-15% of Gram-negative blood culture isolates are not identified byVERIGENE® .While a number of these isolates are species often considered as contaminants in blood cultures (such as species of the genera *Corynebacterium*, *Rothia*, *Micrococcus, Bacillus* and *Gemella*), a significant number of true pathogens remain unidentified (*Morganella*, *Pasturella*, *Serratia*, *Stenotrophomonas*, certain enterococci and some *Acinetobacter* species).In these cases, important clinical decisions regarding antibiotic treatment still need to be made despite the lack of early identification of the pathogens. In this study, there was a hospital mortality rate of 10.0%. Of the Gram-negative bacteria not identified byVERIGENE® , the vast majority were aerobic Gram-negative bacilli,*Serratia* was the most common Gramnegative genus not identified (this species is targeted in the new generations ofVERIGENE test®), followed *byBacteroides*.For Gram-positive cocci, staphylococci and streptococci were the most common organisms not identified by VERIGENE® despite the existence of genus-level targets for *Streptococcus* and *Staphylococcus* included in the microarray, indicating the inherent difficulty of making a highly sensitive probe to target such large and variable genera. Antibiotic use in the 24 hours before and 24 hours after VERIGENE® results were made available in the electronic medical record for Gram-negative and Gram-positive cocci was as follows:almost all patients (98.8%) received antibiotics within 24 hours of the isolation of unidentified Gram-negative organisms from a blood culture (compared with 89% before a positive blood culture), with a significant increase in the use of betalactam antibiotics (from 69.5% to 86.6%), due to a 13.4% increase in the use of piperacillin-tazobactam (from 29.3% to 42.7%). The use of several antipseudomonas antibiotics (piperacillin-tazobactam, cefepime, levofloxacin and meropenem) increased despite the absence of *P.aeruginosa* among the isolates included in this study, and there was a corresponding decrease in the use of ceftriaxone despite similar levels of resistance to ceftriaxone and piperacillin-tazobactam.VERIGENE® performs well when it comes to identifying the isolates it is designed to target, with only a small fraction of the isolates that should have been identified being missed, and many of these were cases where the bacteria were found in polymicrobial cultures**[82]**.

2.3.2.9. Unyvero Blood Culture

Burrack-Lange et al. evaluated the performance of the fully automated *Unyvero Blood Culture* (BCU) system (Curetis GmbH).The *Unyvero* BCU system *is* a cartridge-based molecular diagnostic platform for the detection of bacteria and fungi, as well as antibiotic resistance markers. The application includes 36 analytes covering more than 50 pathogens and 16 antibiotic resistance markers for detection from positive blood culture bottles in around 5 hours. After transferring a sample from the blood culture vial into a tube, lysis is performed in the instrument for 30 min, then the tube is transferred to a cartridge in the presence of a *Master* mix tube *and* the cartridge is analysed by the *Unyvero* system. In this study, the panel had an overall sensitivity of 96.8% and specificity of 99.8%, accurately identified 95% of species and detected a large number of resistance markers, including *mecA*, aminoglycoside resistances, *vanB* and blaCTX-M. All Gram-negative bacteria were identified by this method. Identification of pathogens and resistance markers was available approximately 34 hours faster than the standard method. As this method does not require an initial Gram stain and the workflow is quick and easy, turnaround time could be further reduced if the system is used as close to the patient as possible. UnyveroBC *appears to* be adequate for detecting polymicrobial infections**[83]**.

2.3.2.10. MinION

Ashikawa et al.sought to evaluate the MinION sequencer (Oxford Nanopore Technologies [ONT], Oxford, UK), for the identification of the most common pathogen genes in positive blood culture bottles.The *MinION* sequencer is a small handheld device operating from a laptop connected via *Universal Serial Bus*(USB),maintenance-free, low-cost and multiplex attractive for most clinical situations. The 16S rRNA and ITS1-5.8S-ITS2 genes were amplified by PCR reactions with barcoded primers using nine clinical isolates obtained from positive vials and 11 type strains, including five types of *Candida* species. The barcoded amplicons were mixed and multiplex sequencing with the MinION sequencer was performed. The main innovations of the MinION platform include significantly reduced initial and operational costs and real-time sequencing.The bacterial and fungal species identified by MinION sequencing, based on the highest homology of reference sequences from NCBI gene databases, were consistent with MALDI-TOF results, except *that Streptococcus pyogenes* was identified as *Streptococcus dysgalactiae* and *E. coli* was identified as *Shigella dysenteriae*. However, these errors were caused by the limitations of 16S rRNA sequencing and not by a specific MinION sequencing error, so it is important to bear in mind that this protocol still has difficulty in accurately assigning specific isolates to a particular species. This technique allows the identification of pathogenic bacteria and *Candida* spp, which are the main pathogens responsible for high morbidity and mortality, and which can be identified within 10 minutes of sequencing.however, the low amount of pathogenic DNA and the presence of various PCR inhibitors in the blood culture bottles significantly influence the PCR amplification results. This method could not treat mixed polymicrobial infections or contamination in the culture flasks. In addition, resistance associated with gene mutations or single nucleotide polymorphisms, such as fluoroquinolone resistance, is difficult to determine in MinION **[84]**.

2.3.3. Fluorescence in situ hybridization

Many obligate or facultative intracellular bacteria pose a critical diagnostic problem in clinical microbiology due to their fastidious growth or lack of growth in conventional culture media. Molecular diagnosis is based on the analysis and identification of nucleic acids (DNA and RNA). In the field of infectiology, it combines laboratory medicine with molecular genetics technology to identify infectious pathogens. Fluorescence in situ hybridisation (FISH) is used for the detection and localisation of nucleotide sequences in various samples while preserving cellular integrity. FISH techniques have been applied in cases of blood culture negative endocarditis, respiratory infections, gastrointestinal diseases, mycobacterial infections, highly pathogenic micro-organisms and other diseases with fastidious bacteria such as spirochetes**[85]**.

FISH allows the identification of target microbial organisms (bacteria, yeasts and protozoa) at genus or species level by binding the specific oligonucleotide probe, generally a short probe of 18 to 25 base pairs, to ribosomal RNA for subsequent analysis under a fluorescence microscope. FISH probes for various indications are deposited in the "probeBase" database.

After fixation on conventional slides to preserve cell morphology for tissue staining and to ensure retention of target sequences, a sample preparation step is required (e.g. preceded by target-specific permeabilisation steps to allow standardised uptake of the probe into the cells), hybridisation is performed, in which the specific probe binds to the complementary ribosomal RNA of the target organism, this step is influenced by a variety of factors. Among the most important to take into account when designing the probe are the length and GC content of the probe, as GC bonds are stronger than A-T(U) bonds, the binding strength also depends on the hybridization temperature, the salt concentration, the concentration of the denaturing agent formamide in the hybridization and the wash buffer. Excess probes are removed in a wash step **[86]**. **Figure 6** describes the principle of the FISH technique

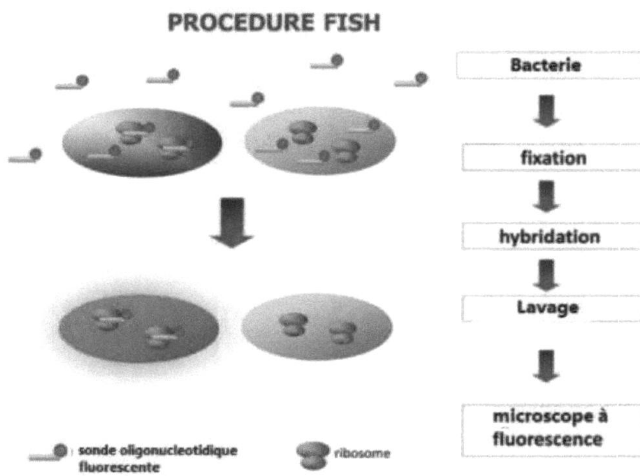

Figure 6: Fluorescence in situ hybridization steps [86].

Figure 7 shows an example of microscopic observation of a Staphylococcus in a blood culture.

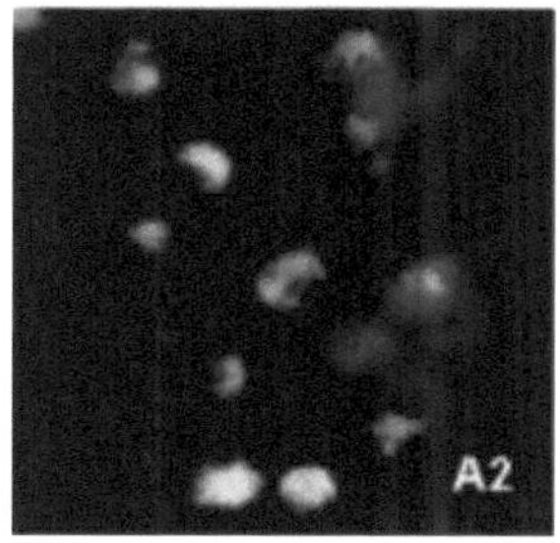

Figure 7: FISH microscopic photograph of a blood sample from a septic patient showing *Staphylococcus* spp. [87].

Zhang et al **[88]** used this method to detect the presence of *S. aureus* in blood culture bottles. In this study, three molecular beacon (MB) probes were used: MB1, MB2, MB3 (an MB is a double-labelled stem-loop oligonucleotide probe that forms a hairpin structure at the 5' and 3' ends, under natural conditions, fluorescence resonance energy transfer occurs in MBs and fluorescence is almost completely suppressed, With the existence of the target sequence, the complementary region of the hairpin structure would be opened, and MB fluorescence is restored) were designed and synthesised to target the fragment of the gene encoding *S. aureus* 16S rRNA. *aureus*, the MB-FISH system was optimised and the specificity of this method in detecting *S. aureus* was assessed, conditions such as deionised formamide concentration, ionic strength and temperature were optimised by monitoring the fluorescence intensity of MB1 in the presence or absence of its target sequence B1.When the temperature was below 50°C, the fluorescence intensity of MB1 without a target sequence in solution was kept low (close to background). When the temperature was between 50°C and 65°C, the fluorescence intensity of MB1 without a target sequence in the solution gradually increased. When the fluorescence intensity of MB1 hybridised with its target sequence B1 decreased, the hairpin structure of MB1 gradually opened up. When the temperature exceeded 65°C, the fluorescence intensity was fundamentally unchanged, indicating that the hairpin structure of MB1 had been completely opened. Therefore, the hairpin structure of MB1 can be opened and closed as a function of temperature. Zhang et al.also evaluated the concentration of deionised formamide, when it was 25% to 40%, the fluorescence intensity of MB1 hybridised with target sequence B1 was the highest, when it was > 40%, the fluorescence intensity of MB1 without target sequence B1 in the reaction system increased significantly. Therefore, when the concentration of deionised formamide was 25% to 40%, hybridisation of MB1 can achieve higher fluorescence intensity and lower background that satisfies MB design requirements.

This detection method has demonstrated high specificity and sensitivity, and is very consistent with the traditional culture method widely used in clinical practice. It also provides a feasible and less expensive framework for the design and evaluation of MB probes for the rapid detection of pathogenic bacteria**[88]**.

Denise et al. studied the AdvanDx QuickFISH test to differentiate true *S.aureus* infection from SCN contamination of blood culture as quickly as possible in order to stop the unwarranted administration of vancomycin. The QuickFISH test is based on fluorescence in situ hybridisation (FISH) using labelled peptide nucleic acid (PNA) probes that target *S.aureus* 16S ribosomal RNA or SCN. The use of PNA FISH molecular diagnostic tests to differentiate *S.aureus* from SCN in blood cultures is associated with reductions in length of stay, antibiotic use, mortality and cost. The test requires minimal training, is simple to perform, has minimal initial set-up costs and reagent costs, and does not significantly affect the workflow of a microbiology laboratory. With a turnaround time of around 5 min, the speed and robustness of QuickFISH allows near-simultaneous reporting of Gram stain results and pathogen identification within 30 minutes, offering critical time savings in formulating patient management decisions. Internal validation of the QuickFISH test has demonstrated 100% agreement between conventional testing and QuickFISH results **[89]**.

Helena et al evaluated the performance of two commercial tests, QuickFISH (OpGen) and MALDI Sepsityper™ (Bruker Daltonics) for early and accurate identification of micro-organisms directly from positive blood cultures. The QuickFISH test saved time and showed a high concordance with the reference method of 94.8%.The sensitivities of the *Staphylococcus* and *Enterococcus* QuickFISH kits were satisfactory, and QuickFISH was able to correctly differentiate SCN from *S. aureus with* a sensitivity of 94.8%. The QuickFISH tests are capable of detecting up to a maximum of three different species per test **[90]**.

Despite the possibility of identifying the most widespread pathogens, several systems are unable to provide reliable identification for rare bacteria or only allow identification at genus level due to the limit on the number of probes included. The PNA-FISH® was the first commercially available system based on FISH technology. It allows the identification of Staphylococci, CNS, Enterococci, *E. coli*, *K. pneumoniae*, *P.aeruginosa*, and 5 species of *Candida* in 2.5 hours. Result times have been reduced to 25 minutes with the development of the QuickFISH test**[60]**.

2.3.3.1. Accelerate Pheno System™

Accelerate Pheno system™ (APS)is a fully automated system that combines rapid pathogen identification and antibiotic susceptibility testing directly from the positive blood culture flask. Cells are stained with a selected panel of FISH probes to identify the microbial species most commonly responsible for sepsis. The cells are then immobilised in agarose layers impregnated with multiple concentrations of appropriate antibiotics, and cell growth is monitored by automated accelerated microscopy to calculate MIC values against a predictive database for susceptible, intermediate and resistant phenotypes based on current *Clinical & Laboratory Standard Institute* (CLSI) or European Committee on Antibiotic Susceptibility Testing (EUCAST) guidelines. The behaviour of the organism is mapped against a predictive database for susceptible, intermediate and resistant phenotypes. The APS system has recently been introduced into microbiology laboratories to obtain the identification of a panel of bacteria and yeasts in 1 hour from the positive blood culture bottle. Antibiotic susceptibility test results with MICs can also be available within 6 to 7 hours of a positive blood culture, for a number of pathogenic agents commonly responsible for bacteremia. Cenci et al. evaluated the potential positive clinical role of PSA in the management and treatment of affected patients in a review including several studies. For the identification of pathogens, overall agreement between PSA and the various techniques (MALDI-TOF, MicroScan WalkAway) ranged from 87.9 to 100%, with the highest performance observed for enterobacteria, with 95% correct identification, followed by staphylococci, with 91% correct identification, and enterococci, with 91% correct identification. Studies comparing APS with different commercial methods found APS sensitivity ranging from 91.5 to 97.5%, specificity from 98.9 to 99.6%, sensitivity ranging from 95.6 to 96.7% for Gram-positive bacteria, from 95.3 to 98.5% for Gram-negative bacteria, and from 97.9 to 100% for *Candida albicans* and *Candida glabrata*.Major identification problems were found for streptococci (a-hemolytic streptococci (*Streptococcus mitis* group and *Streptococcus sanguinis* group) are reported as *S. pneumoniae,* while, using the new APS 1.2 version, out-of-panel ʙ-hemolytic streptococci (*S. pyogenes* and *S. dysgalactiae*) were reported as SCN and *Enterobacter* spp. For the accuracy of antibiotic susceptibility testing, PSA performance was studied in terms of categorical agreement (CA), defined as the percentage of total test results with the same categorical interpretation of the reference method, essential agreement (EA), the percentage of total test results at ±1 dilution to 1/2 of the reference result, percentage of very major errors (VME) of resistant isolates tested susceptible by APS, major errors (ME), susceptible isolates tested resistant by APS and minor errors (ME) intermediate result and the other susceptible or resistant.According to CLSI criteria, the performance of the commercial antibiogram must be > 90% AC and > 90% AE, acceptable EM and ETM rates are generally 3% of the susceptible and resistant isolates tested, and acceptable Em rates are 10%. In most studies overall AC and AE were both above 91%, for Gram-positive bacteria AC and AE were 97.9 and 97.6%, respectively, and for Gram-negative bacteria AC and AE were 94.3 and 95.4%, respectively.ETM and EM were mainly found with ampicillin-sulbactam, ceftazidime, piperacillin/tazobactam.APS can reduce the time required to switch from empirical broad-spectrum antibiotics to definitive treatment. Implementation of APS in the clinical management of bacteremias could hypothetically reduce the time to definitive treatment by at least 25.4 h. In particular, APS has been shown to reduce the time to definitive treatment by 37.8 h for carbapenem-resistant enterobacteria, 35.9 h for non-fermentative bacteria and 16.7 h for ESBL-producing bacteria.

The inability of the system to specifically identify *S. pneumoniae* and its possible resistance to penicillin, as well as the inability to detect species other than *C. albicans* and *C. glabrata* and to provide antifungal sensitivity results, are among the limitations of APS. *P. aeruginosa* has emerged as the most difficult pathogen for susceptibility testing, and piperacillin-tazobactam and colistin as the antibiotics most prone to susceptibility errors.The identification of pathogens responsible for polymicrobial bacteremia needs to be optimised, and the performance and utility of PSA in polymicrobial infections needs to be further investigated. Given the high cost of the test, it is recommended that Gram staining, and identification of the isolate by MALDI-TOF, be performed before deciding to use this technique [91].

In this study, Starr et al. evaluated the performance of a version of the APS system using positive clinical blood cultures incubated in resin-containing vials. These vials did not interfere with assay performance. The mean identification time was reduced by 24.9 ± 6.9 h and the susceptibility test by 36.7 ± 18.9 h compared with the standard method. The antibiogram had an AE of 96.5% and an AC of 94.6% with someEM andETM.Some limitations in the FDA pre-approved version of the software were observed. In this study, there were 5% of samples where the BacT/ALERT® system marked a vial as positive but there were too few cells for analysis by APS (described as an identification failure).However, the identification failure rate was resolved in a software update.In the case of an identification failure, a report would be generated indicating the error about 90 min after loading and the run would terminate at that time.For the antibiogram, which takes about 7 h to complete, failures could have a greater impact**[92]**.

Descours et al. evaluated the performance of the APS software compared with routine techniques (MALDI-TOF and VITEK® 2). Complete concordance between APS and MALDI-TOF for identification was observed for 96.2% of samples, and was 99% for monomicrobial samples compared with 40% for polymicrobial samples. The CA for antibiotic susceptibility testing was 93.7%. It was particularly low for beta-lactams (cefepime 84.4%, piperacillin-tazobactam 86.5%, ceftazidime 87.6%) and *P. aeruginosa* (71.9%; with cefepime 33.3%, piperacillin-tazobactam 77.8%, ceftazidime 0%). Analysis of discordant results showed that APS performed poorly for beta-lactams in enterobacteria and poorly in *P. aeruginosa*, with 8.2% of APS-initiated susceptibility tests failing compared with only 1.1% of VITEK-initiated susceptibility tests 2.®

APS performed relatively well for aminoglycoside testing, with poorer performance for *P. aeruginosa* than for enterobacteria, and significantly improved the turnaround time for Gram-negative bacteria diagnosis compared with conventional routine techniques. While most previous studies fully support its implementation in a routine clinical microbiology laboratory workflow, some authors point out that the system still needs improvements both in the identification of polymicrobial samples and in susceptibility testing of Gram-negative bacteria, particularly for betalactam antibiotics**[93]**.

2.3.4. *Loop-mediated isothermal amplification*

Rodel et al. evaluated the *loop-mediated isothermal* amplification test to rapidly identify pathogens and resistance genes directly from positive blood cultures. The commercially available *eazyplex*® MRSA test detects *S. aureus*, *S. epidermidis*, *mecA* and *mecC*. The other two tests [Gram-positive (GP) and Gram-negative (GN)] were developed for the detection of streptococci, enterococci, *vanA*, *vanB*, Pseudomonasspp, *Enterobacteriaceae* and the blaCTX-M family. Test results were obtained in 30 minutes, including sample preparation. Amplification was measured by real-time fluorescence detection. The threshold time for fluorescence intensity values ranged from 6.25 to 13.75 min. With the exception of *K. pneumoniae*, false positive signals were generally detected later than true positive signals (mean value=15.25 min) **[94]**.

The specificity and sensitivity of the tests varied according to the target. Positive results were obtained for 87.7% of vials compared with standard routine diagnosis. The concordance of positive test results for resistance genes with antibiotic sensitivity tests was 100%. Limitations of this study include the limited panel of pathogens tested. The Eazyplex® strip format only allows a maximum of seven target genes and one inhibition control. This study shows that these tests are rapid, easy-to-handle and cost-effective tools for rapid testing of blood culture bottles to speed up result reporting and improve antibiotic stewardship for patients with bacteremia **[94]**.

2.3.5. BinaxNow technique

Heraud et al. evaluated the accuracy of two immunochromatographic tests, BinaxNow *S. aureus* for the identification of *S.aureus* and BinaxNowPBP2 for the determination of methicillin resistance, applied directly to positive blood culture bottles. Sensitivity and specificity for the identification of *S. aureus* and the determination of methicillin resistance were 94% and 87%, and 100% and 100%, respectively, with less than 30 min performance time.these methods perform well in the rapid detection of *S. aureus* /SCN and methicillin sensitivity or resistance. A positive BinaxNow *S. aureus* result could be reported to clinicians as '*S. aureus*', whereas a negative result should only be reported as 'test negative', in association with direct blood culture results. A negative test could be due to CNS as well as other Gram-positive bacteria in clusters; however, this test has not been developed for SCN strains. Molecular tests have also been widely evaluated**[95]**.

The GeneOhm real-time quantitative multiplex PCR assay showed better sensitivity and specificity for the detection of *S. aureus* than BinaxNow *S. aureus*, while BinaxNowPBP2a showed better results for the detection of MRSA than GeneOhm.

The fully automated Xpert MRSA/SA test is better at detecting *S. aureus* but not MRSA than BinaxNow *S. aureus* and BinaxNowPBP2a.

MALDI-TOF can identify *S. aureus* and SCN directly from the blood culture bottle. However, to date, MALDI-TOF cannot detect methicillin resistance. Therefore, an alternative method may be the combination of MALDI-TOF for the identification of *S. aureus* and a BinaxNowPBP2a test to determine methicillin resistance **[95]**.

2.3.6. Identification techniques only

2.3.6.1. Liquid chromatography and tandem mass spectrometry

Berendsen et al. have evaluated a method based on liquid chromatography and tandem mass spectrometry that can identify highly pathogenic bacteria directly from positive blood culture bottles, enabling rapid identification of micro-organisms on the basis of proteome analysis. Sample preparation begins with a lysis step using 2.5% saponin and trypsin digestion. A spectrum is generated. The eluted peptides are analysed using tandem mass spectrometry. The database contains 442 entries, representing 429 bacterial species and 9 yeast species. The high number of different species currently included in the database increases the chances of rapid diagnosis in the case of uncommon pathogens and, unlike targeted methods (e.g. PCR or immune tests), is unaffected by mutations in the pathogen genome, thereby increasing the reliability of identification.The time required to identify a micro-organism is around eight hours per sample. The main limiting factor in this

This method may initially be used for difficult priority samples. For example, it could be useful for blood culture bottles collected from patients or travellers with unpredictable causes of sepsis[96].

2.3.6.2. Gas chromatography coupled to an ion mobility spectrometer

Drees et al. sought to investigate the ability to detect volatile organic compounds of microbial origin mVOC using a gas chromatograph coupled to a high-sensitivity ion mobility spectrometer for the recognition of bacterial growth and bacterial differentiation in positive haemocultures.the method offers high sensitivity with detection limits ranging from nanograms per litre to picograms per litre in a few seconds of analysis. The molecules are ionised and separated in an electric field by collisions with molecules of the drift gas by their mass, size and shape. Headspace samples from cultures of *E.coli, S. aureus,* and *P. aeruginosa* were studied every hour and determined when differentiation between bacteria was possible. mVOC-specific signals were found in the headspace for growing flasks of all three bacterial species. Analyses using this technique enabled faster recognition of bacterial growth than the CO_2-sensitive colourimetric indicator of blood culture bottles. In this study, 3-methyl-butanal was identified as a mVOC specific to *S. aureus*.the acetone observed was defined as a mVOC bound to *E. coli*.none of the mVOCs described as specific to *P. aeruginosa* could be identified.however, the specific production of mVOCs by a bacterium is highly dependent on the culture media.Differentiation between the three species studied was possible after 6 h of incubation with high reliability in principal component analysis.this technique can be considered a useful tool for rapid identification of bacteria in clinical routine once a complete database of mVOC and mode characteristics for all relevant pathogens has been implemented**[97]**.

2.3.6.3 *ImmuLex™*

Altun et al. have evaluated a new latex agglutination test. The rapid identification of *S. pneumoniae* in blood culture bottles by ImmuLex™ S. pneumoniae Omni (Statens Serum Institut, Denmark). They compared performance with the Slidex® pneumo-Kit (bioMerieux, France) and the Wellcogen™ S. pneumoniae latex agglutination test

(Remel, UK), as well as the BinaxNOW® S. pneumoniae antigen test (Alere, USA).ImmuLex sensitivity was 99.6%, similar to the other tests.Valid test results were recorded in all cases for ImmuLex and BinaxNOW and in 88.5% and 94.1% of cases for Slidex and Wellcogen.On the basis of vials positive for non-pneumococcal pathogens, the specificity of ImmuLex was 82.6%, compared with 97.6% for Slidex and 85.4% for Wellcogen. Of the BC vials positive for a-hemolytic streptococci, ImmuLex was positive in 17.9%, Slidex in 3.4%, Wellcogen in 17.2% and BinaxNOW in 37.3%. The ImmuLex test provides a valid and sensitive technique for the rapid detection of *S. pneumoniae* in blood culture bottles, similar to the other methods compared. However, specificity was suboptimal, as the test can cross-react with other Gram-positive bacteria, viridans streptococci, *S. sanguinis*, *S. epidermidis/Enterococcus* species (in combination) and *Pseudomonas* species.False-positive results from the ImmuLex test have previously only been described on vials positive for *Streptococcus* C and *P. aeruginosa* in the time course of the test. Due to a high rate of false positive test results on vials positive for a-hemolytic streptococci, the test was less specific than the Slidex test but similar to the Wellcogen test**[98]**.

2.3.7. Antibiotic sensitivity testing techniques only 2.3.7.1. Gelose diffusion technique

A method for rapid determination of antibiotic susceptibility of bacteria directly from positive blood culture bottles based on gelose diffusion has been developed. It includes a number of modifications compared with the classic method: inoculum, shorter incubation time, reading instructions and new specific critical concentrations. This method is only valid for a maximum of 8 hours incubation in Petri dishes. If incubation is to be extended beyond 8 hours, the standard gelose diffusion method should be used. After 4 to 8 hours incubation, growth on Muller-Hinton gelose will often appear less distinct compared with standard incubation (16-20 hours). This technique is valid for *E. coli, K. pneumoniae, A. baumannii, S. aureus, E. faecalis, E. faecium*, *P. aeruginosa and S. pneumoniae***[17]**.

2.3.7.1. *Real-time video imaging*

The gelose diffusion method has long been considered the standard technique for antibiotic susceptibility testing due to its simplicity, reproducibility and low cost compared to commercial automated microdilution systems which are generally faster but less sensitive for detecting important resistance mechanisms. Le Page et al. measured the diameters of the zones of inhibition around antibiotics for Gram-negative bacteria using an Advencis Bio-System instrument consisting of an integrated incubator capable of incubating Petri dishes and obtaining images directly using dedicated software for image analysis using real-time video imaging. The sensitivity of Gram-negative bacteria to imipenemia was detectable as early as 3.5 h (mean time 3.7 ± 0.45 h), while carbapenemase-producing Gram-negative bacteria could be correctly classified as early as 3eme h (mean time 4.2 ± 0.8 h) of incubation. Similarly, the champagne cork appearance characteristic of ESBLs could be detected by the 3.5 h system. This new and innovative technology is of interest to microbiologists who wish to continue using the diffusion method for antibiotic susceptibility testing **[99]**.

2.3.7.2. FASTinov®

Costa de Oliveiraet al**[100]** evaluated the performance of the new FASTinov® kit for antibiotic susceptibility testing of Gram-negative bacilli directly from positive blood cultures, using the routine VITEK2 method, the FASTinov® kit and the reference microdilution. The FASTinov® kit for Gram-negative bacteria was used after observation of the Gram stain (immediately after the blood culture was reported positive). The FASTinov® *gramneg* kit comprises a microplate containing a panel of lyophilised antibiotics, a specific fluorochrome and dedicated software (which allows selection of EUCAST and CLSI classification rules). A flow cytometric analysis was then performed on a flow cytometer and the result was automatically classified as sensitive (S), intermediate (I) or resistant (R). The mean time to result for the routine test (VITEK2) was a minimum of 24 hours compared with a 2-hour response time for the Fastinov test® , allowing earlier switching to a narrow-spectrum targeted therapy, thereby reducing selection pressure and minimising exposure to inappropriate therapy and its adverse effects. The overall CA between the test and the reference method was 98%, with Em, EM and ETM of 1.03, 1.59 and 1.06%, respectively. According to the FDA, the rate of EM must be <1.5%.ETMs obtained with the kit were detected with amoxicillin clavulanic acid (2.00%), gentamicin (2.78%) and meropenem (6.25%). The kit correctly detects resistance phenotypes to amikacin, cefotaxime, ceftazidime, ciprofloxacin, colistin and piperacillin-tazobactam, providing no false sensitivity results. For the piperacillin-tazobactam combination, only one EM (1.15%) and one Em (0.78%) were observed with the FASTinov® *gramneg* kit. The FASTinov® *gramneg* kit also provided correct detection of ESBL-producing strains (AC 100%). The ability of FASTinov® gramnega to promote optimal antibiotic use deserves to be tested in several laboratories. Additional benefits could include reduced hospital length of stay, lower antibiotic consumption and better infection control **[100]**.

2.3.7.4 Uro4 HB&LTM system

Athamna et al **[101]** evaluated the performance of the Uro4 HB&LTM system (Alifax SpA) in the detection of ESBL/AmpC-producing enterobacteria compared with the VITEK 2 resistance test.

The principle of the method is based on the scattering of laser light to follow the culture in liquid medium for 6.5 hours of incubation. A laser beam is passed through glass vials containing theBLSE/AmpC reagent (a mixture of antibiotics) to which the bacterial suspension is added. The light scatter signals are detected and displayed in the form of growth curves and a count evaluation is obtained by algorithmic calculation.there was a 91.3% concordance with the VITEK2 system, with a sensitivity of 85%, a specificity of 95.2%.the relatively low sensitivity, compared to the specificity, is related to the results *ofProteus mirabilis*, in which the system did not detect ESBL-producing *P. mirabilis* (false negatives).The combination of MALDI-TOF technology with the Uro4 HB&LTM system makes it possible to identify the species of bacterium responsible and the partial antibiotic resistance profile, such as the ability to produce ESBLs. However, the test does not determine the MICs of the antibiotics tested and does not distinguish between ESBL and AmpC mechanisms. ESBLs in these bacteria, such as CTX-M, SHV and TEM**[101]**.

2.3.7.5. Fluorogenic probe

Kim et al **[102]** have developed a rapid fluorogenic assay using a carbapenem-based fluorogenic probe conjugated to hydroxybenzyloxy umbelliferone. **Figure 8** shows the hydrolysis of the specific probe by carbapenemase.

The fluorescent signal from carbapenemase-producing enterobacteria was significantly higher than that from non-producers **[102]**.

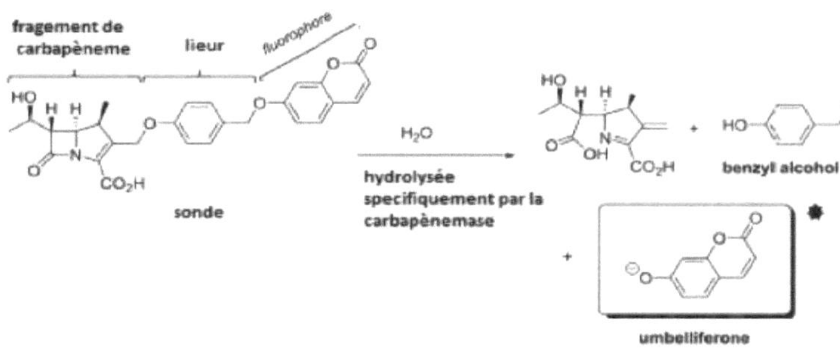

Figure 8: Fluorogenic probe and its hydrolysis by carbapenemase [from 102].

2.3.7.6 *Direct & Rapid Antimicrobial Susceptibility Test* System

Rosselin et al **[103]** tested the Direct & Rapid Antimicrobial Susceptibility Test (dRAST) system, a rapid antibiogram system based on microscopic imaging. **Figure 9** describes the different stages of this analysis. The overall CAs for enterobacteria, non-fermentative Gram-negative bacteria, staphylococci and enterococci were 95.1%, 91.2%, 93.4% and 94.5%, respectively. The percentage of TME was less than 4% for all groups except staphylococci, which had a TME rate of 7%. The median time to result was 6.7 hours. Most of the EM and ETM observed with *Staphylococcus* species were due to oxacillin. Levofloxacin showed high levels of ME or TME with Staphylococcusspp, *Enterococcus* spp and non-fermentative Gram-negatives. Therefore, systematic verification of levofloxacin by another method would be required before reporting the final results. For enterobacteria, most of the EMs and ETMs were obtained with в- lactamins. For non-fermentative Gram-negatives, TMEs were obtained with levofloxacin and MEs with cefepime, piperacillin-tazobactam and trimethoprim-sulfamethoxazole. This test is more expensive than conventional sensitivity tests **[103]**.

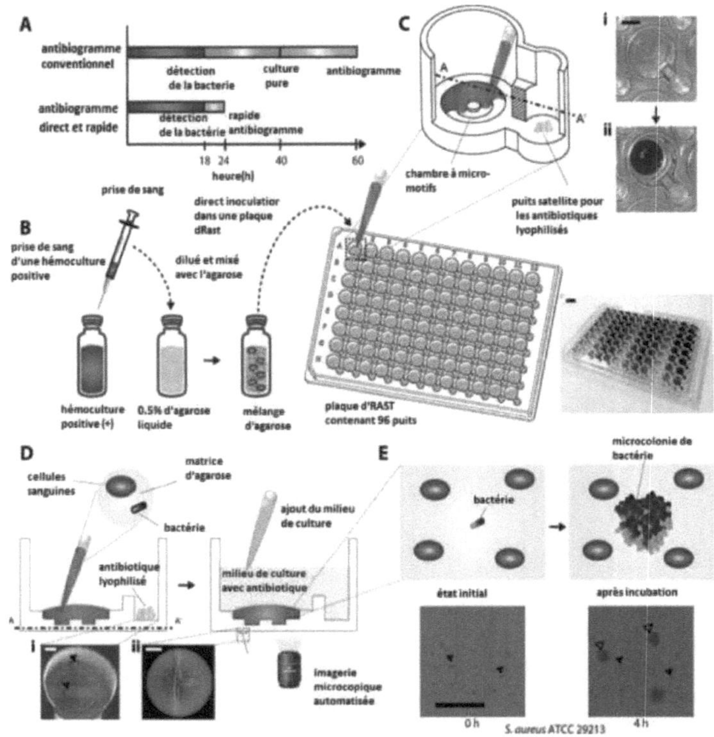

Figure 9: Device for direct detection of antibiotic resistance in positive blood culture bottles using the dRAST system [104].

(A) Conventional antibiotic susceptibility testing requires three separate culture steps: blood culture for detection of infection, subculture for separation of bacteria from blood cells and the actual susceptibility test; the total time to result is approximately 60 hours. The dRAST system does not require a subculture process and results can be obtained in six hours. The total time to obtain the dRAST result is less than 24 hours. (B) Preparation of a dRAST test from a positive vial. An aliquot is diluted and mixed with liquid agarose. The mixture is inoculated into a dRAST plate consisting of 96 test wells containing various antibiotics at different concentrations. (C) Detailed structure of a dRAST plate well. Each test well consists of a micropatterned radial chamber for moulding the agarose matrix and a satellite well for freeze-dried antibiotics. The inset images show the micropatterned chamber i) before loading of the agarose mix and ii) after loading of the agarose mix. (D) Experimental process for the detection of bacteria in blood. A sample of broth from a positive vial is mixed with agarose and loaded into the plate loading chamber. The lyophilised antibiotics are rehydrated by adding culture medium (Mueller Hinton broth adjusted for Ca^{2+}). Automated microscopic imaging is used to detect bacterial growth. Medallion scanning electron microscope images show (i) lyophilised antibiotics in the satellite well and (ii) the focus mark at the bottom of the chip for automated imaging. (E) In the initial state, bacteria were not detectable using a 20x magnification lens; only blood cells were detectable. After four hours of incubation, in the case of antibiotic resistance, a single bacterium divides and forms a micro-colony. The solid triangle represents a blood cell. The empty triangle represents a bacterial micro-colony. The scale bar represents 10 mm in (B), 3 mm in (C), 300 pm in (D) and 100 pm in (E).

2.3.7.7 *Scattered Light Integrated Collection*

Falconer et al **[105]** conducted a clinical study to evaluate the performance of a *Scattered Light Integrated Collection* device as *an* accurate and rapid antibiotic susceptibility test for Gram-negative bacteria that can monitor discrete changes in bacterial cells as a function of the total light scatter detected. Differentiation between cell response of either growth, growth inhibition or cell death is possible due to the distinct change in light scattering properties. **Figure 10** describes how this method works.

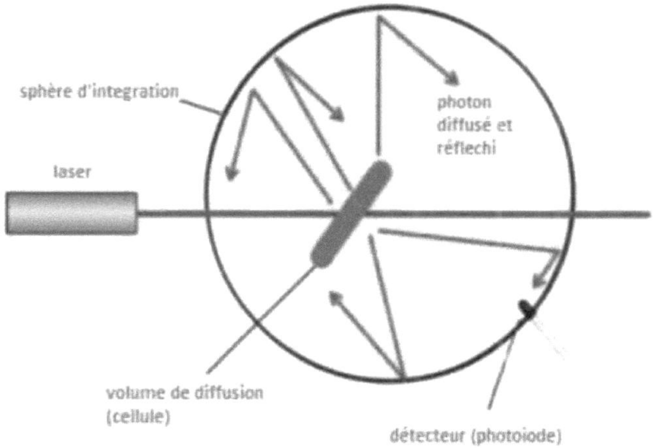

Figure 10: Principle of the *Scattered Light Integrated Collection* method [105].

Scattered Light Integrated Collectionse consists of a laser, a light collector and innovative electronics. This technique detects the total diffusion of light, giving a high detection sensitivity of 100 cells/mL.

This technique provided consistent and accurate susceptibility test results in 2 h with low error rates. This method accelerated the time to obtain an antibiogram result compared with workflow **[105]**.

2.3.7.8 Colorimetric tests

Seco et al. used 2 colorimetric techniques which are Carba-NP andBlue- Carba initially described for use with bacterial colonies grown on solid media. They used the bacterial cell pellet to identify carbapenemase-producing enterobacteria with species identification by MALDI-TOF. Carbapenemase-producing enterobacteria were revealed by detecting acid production during hydrolysis of imipenem. Handling time was 13 min. The technique does not require the use of ammonium chloride for the lysis of bacterial cells, but polybial bacteremia can affect the performance of the analysis **[106]**.

2.3.8. Qvella's FAST™ system

Grinberg et al. evaluated a new device, the FAST™ system fromQvella, which concentrates and purifies microbial biomass, from positive blood culture flasks with the FAST-PBC Prep™ cartridge, producing a pure, concentrated suspension in about 30 minutes, as shown in **Figure 11** and requires about 2 minutes of handling time. The advantage of the FAST™ system is its simplicity in purification directly from the positive blood culture vial. The resulting suspension can then be used for identification and antibiotic susceptibility testing **[107,108]**.

transfer! of the treated sample from the cartridge recover the liquid colony

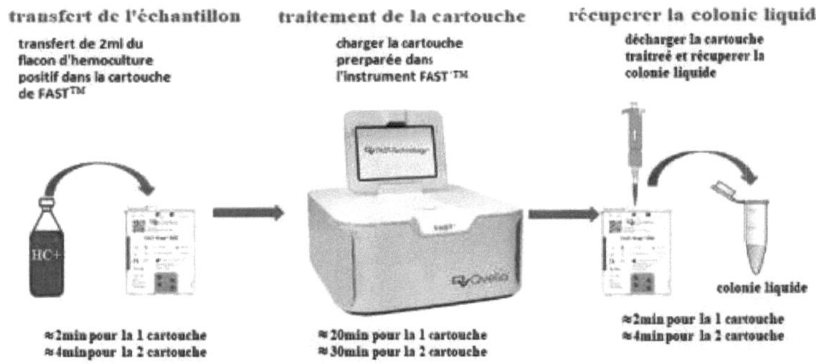

Figure 11: Steps in the preparation of the purified suspension using the technique FAST-PBC Prep™ [108].

CONCLUSION

New rapid molecular and phenotypic tests for identification and antibiotic susceptibility testing have recently been introduced into modern clinical bacteriology laboratories. They have demonstrated their potential to positively influence patient outcomes and antibiotic stewardship interventions, by reducing both the time required for effective treatment and the overuse of new antibiotics.Among these tests, those that have been most evaluated are based on MALDI-TOF, PCR and FISH techniques. Their real impact on therapeutic decisions and the prognosis of bacteremia is still fragmentary. Biologists and clinicians need to be aware of the rapid pace of new developments in this field and antibiotic stewardship programmes need to coordinate to ensure that new technologies optimally affect patient care. Knowledge of the performance of the analytical method used is essential to control the reliability of results. An analytical method has limitations that the laboratory needs to be aware of. Evaluation of the cost-effectiveness of some of these tests has shown that their use is often a cost-effective strategy, associated with high therapeutic efficacy and savings in healthcare costs. It is important to conduct a thorough evaluation of the cost-effectiveness of these tests in Tunisia with a view to their introduction into routine diagnostics. Laboratory implementation of these techniques needs to be carefully tailored according to a number of factors, including hospital, laboratory capacity, clinicians and patient population. Further studies of their performance are required, as are further studies elucidating the benefits of these various technologies in combination with an antibiotic stewardship programme to define best practice.

REFERENCES

1. **Huang M, Cai S, Su J.** The pathogenesis of sepsis and potential therapeutic targets. *Int J Mol Sci. 2019;20:5376.*
2. **Woznica E, Inglot M, Woznica R, Lysenko L.** Liver dysfunction in sepsis. *Adv Clin Exp Med. 2018;27:547-52.*
3. **Quiles MG, Boettger BC, Inoue FM, Monteiro J, Santos DW, Ponzio V, et al.** Direct matrix-assisted laser desorption ionization time-of-flight mass spectrometry and real-time PCR in a combined protocol for diagnosis of bloodstream infections: a turnaround time approach. *Braz J Infect Dis. 2019;23:164-72.*
4. **Peker N, Couto N, Sinha B, Rossen JW.** Diagnosis of bloodstream infections from positive blood cultures and directly from blood samples: recent developments in molecular approaches. *Clin Microbiol Infect. 2018;24:944-55.*
5. **College des Universitaires des Maladies Infectieuses et Tropicales (CMIT).** E.Pilly, infectious and tropical diseases. 27[eme] edition. *Paris: Alinea Plus; 2020.*
6. **Ombelet S, Barbe B, Affolabi D, Ronat JB, Lompo P, Lunguya O, et al.** Best practices of blood cultures in low- and middle-income countries. *Front Med. 2019;6:131.*
7. **Giavarina D, Lippi G.** Blood venous sample collection: Recommendations overview and a checklist to improve quality. *Clin Biochem. 2017;50:568-73.*
8. **Lamy B, Dargere S, Arendrup MC, Parienti JJ, Tattevin P.** How to optimize the use of blood cultures for the diagnosis of bloodstream infections? A state-of-the art. *Front Microbiol. 2016;7:697.*
9. **Doern GV, Carroll KC, Diekema DJ, Garey KW, Rupp ME, Weinstein MP, et al.** Practical guidance for clinical microbiology laboratories: a comprehensive update on the problem of blood culture contamination and a discussion of methods for addressing the problem. *Clin Microbiol Rev. 2019;33:e00009-19.*
10. **Huber S, Hetzer B, Crazzolara R, Orth-Holler D.** The correct blood volume for paediatric blood cultures: a conundrum? *Clin Microbiol Infect. 2020;26:168-73.*
11. **Antillon M, Saad NJ, Baker S, Pollard AJ, Pitzer VE.** The relationship between blood sample volume and diagnostic sensitivity of blood culture for typhoid and paratyphoid fever: a systematic review and meta-analysis. *J Infect Dis. 2018;218(suppl_4):S255-67.*
12. **Plato FD, Fontana C, Gherardi G, Privitera GP, Puro V, Rigoli R, et al.** Collection, transport and storage procedures for blood culture specimens in adult patients: recommendations from a board of Italian experts. *Clin Chem Lab Med. 2019;57:1680-9.*
13. **Guna Serrano MR, Larrosa Escartrn N, Marin Arriaza M, Rodriguez D^az JC.** Microbiological diagnosis of bacteraemia and fungaemia: Blood cultures and molecular methods. *Enferm Infecc Microbiol Clin (Engl Ed). 2019;37:335-340.*
14. **Houssaini ZE, Harrar N, Zerouali K, Belabbes H, Elmdaghri N.** Prevalence of coagulase-negative staphylococci in blood cultures at the Centre Hospitalier Universitaire Ibn Rochd de Casablanca. *Pan Afr Med J. 2019;33:193.*

15. Osthoff M, Khanna N, Goldenberger D, Wuscher V, Fluckiger U. Positive blood cultures: interpretation and initial management. *Forum Med Suisse. 2016;16:59-67.*

16. Monteiro AC, Fortaleza CM, Ferreira AM, Cavalcante R de S, Mondelli AL, Bagagli E, et al. Comparison of methods for the identification of microorganisms isolated from blood cultures. *Ann Clin Microbiol Antimicrob. 2016;15:45.*

17. Antibiogram Committee of the French Society of Microbiology. Recommendations 2022. V.1.0 May. *Paris: Societe Franqaise de Microbiologie; 2022.*

18. Tortora GJ, Funke BR, Case CL. Microbiology: an introduction. 13th edition. *London: Pearson Education; 2021.*

19. Pomastowski P, Buszewski B. Complementarity of matrix- and nanostructure-assisted laser desorption/ionization approaches. *Nanomaterials(Basel). 2019;9:260.*

20. Azrad M, Keness Y, Nitzan O, Pastukh N, Tkhawkho L, Freidus V, et al. Cheap and rapid in-house method for direct identification of positive blood cultures by MALDI-TOF MS technology. *BMC Infect Dis. 2019;19:72.*

21. Dingle TC, Butler-Wu SM. MALDI-TOF mass spectrometry for microorganism identification. *Clin Lab Med. 2013;33:589-609.*

22. Tsuchida S, Umemura H, Nakayama T. Current status of matrix-assisted laser desorption/ionization-time-of-flight mass spectrometry (MALDI-TOF MS) in clinical diagnostic microbiology. *Molecules. 2020;25:4775.*

23. Caspar Y, Garnaud C, Raykova M, Bailly S, Bidart M, Maubon D. Superiority of SDS lysis over saponin lysis for direct bacterial identification from positive blood culture bottle by MALDI-TOF MS. *Proteomics Clin Appl. 2017;11(5-6):1600131.*

24. Di Gaudio F, Indelicato S, Indelicato S, Tricoli MR, Stampone G, Bongiorno D. Improvement of a rapid direct blood culture microbial identification protocol using MALDI-TOF MS and performance comparison with SepsiTyper kit. *J Microbiol Methods. 2018;155:1-7.*

25. Jeverica S, Nagy E, Mueller-Premru M, Papst L. Sample preparation method influences direct identification of anaerobic bacteria from positive blood culture bottles using MALDI-TOF MS. *Anaerobe. 2018;54:231-5.*

26. Kayin M, Mert B, Aydemir S, Ozenci V. Comparison of rapid BACpro® II, Sepsityper® kit and in-house preparation methods for direct identification of bacteria from blood cultures by MALDI-TOF MS with and without Sepsityper® module analysis. *Eur J Clin Microbiol Infect Dis. 2019;38:2133-43.*

27. Morgenthaler NG, Kostrzewa M. Rapid identification of pathogens in positive blood culture of patients with sepsis: review and meta-analysis of the performance of the Sepsityper Kit. *Int J Microbiol. 2015;2015:827416.*

28. Scohy A, No6l A, Boeras A, Brassinne L, Laurent T, Rodriguez- Villalobos H, et al. Evaluation of the Bruker® MBT Sepsityper IVD module for the identification of polymicrobial blood cultures with MALDI- TOF MS. *Eur J Clin Microbiol Infect Dis.*

2018;37:2145-52.

29. **Tsuchida S, Murata S, Miyabe A, Satoh M, Takiwaki M, Ashizawa K, et al.** Application of the biocopolymer preparation system, rapid BACpro® II kit, for mass-spectrometry-based bacterial identification from positive blood culture bottles by the MALDI Biotyper system. *J Microbiol Methods. 2018;152:86-91.*

30. **Yonetani S, Ohnishi H, Ohkusu K, Matsumoto T, Watanabe T.** Direct identification of microorganisms from positive blood cultures by MALDI- TOF MS using an in-house saponin method. *Int J Infect Dis. 2016;52:37-42.*

31. **Tanner H, Evans JT, Gossain S, Hussain A.** Evaluation of three sample preparation methods for the direct identification of bacteria in positive blood cultures by MALDI-TOF. *BMC Res Notes. 2017;10:48.*

32. **Barberino MG, Silva M de O, Arraes ACP, Correia LC, Mendes AV.** Direct identification from positive blood broth culture by matrix-assisted laser desorption-ionization time-of-flight mass spectrometry (MALDI-TOF MS). *Braz J Infect Dis. 2017;21:339-42.*

33. **Hu YL, Hsueh SC, Ding GS, Chuang PC, Chen JM, Lu CY, et al.** Applicability of an in-house saponin-based extraction method in Bruker Biotyper matrix-assisted laser desorption/ionization time-of-flight mass spectrometry system for identifying bacterial and fungal species in positively flagged pediatric VersaTREK blood cultures. *J Microbiol Immunol Infect. 2020;53:916-24.*

34. **Yonezawa T, Watari T, Ashizawa K, Hanada D, Yanagiya T, Watanabe N, et al.** Development of an improved rapid BACpro® protocol and a method for direct identification from blood-culture-positive bottles using matrix-assisted laser desorption ionization time-of-flight mass spectrometry. *J Microbiol Methods. 2018;148:138-44.*

35. **Yuan Y, Wang J, Zhang J, Ma B, Gao S, Li Y, et al.** Evaluation of an optimized method to directly identify bacteria from positive blood cultures using MALDI-TOF mass spectrometry. *J Clin Lab Anal. 2020;34:e23119.*

36. **Bazzi AM, Rabaan AA, El Edaily Z, John S, Fawarah MM, Al-Tawfiq JA.** Comparison among four proposed direct blood culture microbial identification methods using MALDI-TOF MS. *J Infect Public Health. 2017;10:308-15.*

37. **Samaranayake WA, Dempsey S, Howard-Jones AR, Outhred AC, Kesson AM.** Rapid direct identification of positive paediatric blood cultures by MALDI-TOF MS technology and its clinical impact in the paediatric hospital setting. *BMC Res Notes. 2020;13:12.*

38. **French K, Evans J, Tanner H, Gossain S, Hussain A.** The clinical impact of rapid, direct MALDI-ToF identification of bacteria from positive blood cultures. *PLoS One. 2016;11:e0169332.*

39. **Lin JF, Ge MC, Liu TP, Chang SC, Lu JJ.** A simple method for rapid microbial identification from positive monomicrobial blood culture bottles through matrix-assisted laser desorption ionization time-of-flight mass spectrometry. *J Microbiol Immunol Infect. 2018;51:659-65.*

40. **McIver CJ, Er N, Mukerjee C, Tokis S, Taylor P.** A simplified and rapid method for the direct identification of microorganisms in positive BacT/ALERT blood culture bottles using MALDI-TOF MS. *Pathology (Phila). 2018;50:676-9.*

41. **Hazelton BJ, Thomas LC, Olma T, Kok J, O'Sullivan M, Chen SC, et al.** Rapid and accurate direct antibiotic susceptibility testing of blood culture broths using MALDI Sepsityper combined with the BD Phoenix automated system. *J Med Microbiol. 2014;63:1590-4.*

42. **March-Rossello GA, Munoz-Moreno MF, Gartfa-Loygorri-Jordan de Urries MC, Bratos-Perez MA.** A differential centrifugation protocol and validation criterion for enhancing mass spectrometry (MALDI-TOF) results in microbial identification using blood culture growth bottles. *Eur J Clin Microbiol Infect Dis. 2013;32:699-704.*

43. **Luethy PM, Johnson JK.** The Use of Matrix-Assisted Laser Desorption/Ionization Time-of-Flight Mass Spectrometry (MALDI-TOF MS) for the identification of pathogens causing sepsis. *J Appl Lab Med. 2019;3:675-85.*

44. **Febbraro F, Rodio DM, Puggioni G, Antonelli G, Pietropaolo V, Trancassini M.** MALDI-TOF MS Versus VITEK® 2: Comparison of Systems for the Identification of Microorganisms Responsible for Bacteremia. *Curr Microbiol. 2016;73:843-50.*

45. **Nomura F, Tsuchida S, Murata S, Satoh M, Matsushita K.** Mass spectrometry-based microbiological testing for blood stream infection. *Clin Proteomics. 2020;17:14.*

46. **Fang C, Zhou Z, Li J, Chen X, Zhou M.** Rapid identification of microorganisms from positive blood cultures in pediatric patients by MALDI-TOF MS: Sepsityper kit versus short-term subculture. *J Microbiol Methods. 2020;172:105894.*

47. **Tsai YW, Lin TC, Chou HY, Hung HY, Tan CK, Wu LC, et al.** Shortening the time of the identification and antimicrobial susceptibility testing on positive blood cultures with MALDI-TOF MS. *Diagnostics. 2021;11:1514.*

48. **Fang YT, Grabsch EA, Leroi MJ, Sherry NL, Ward PB.** Optimising early identification of blood culture isolates by MALDI-TOF and use of multiple target wells. *Pathology (Phila). 2018;50:584-7.*

49. **Zabbe JB, Zanardo L, Megraud F, Bessede E.** MALDI-TOF mass spectrometry for early identification of bacteria grown in blood culture bottles. *J Microbiol Methods. 2015;115:45-6.*

50. **Halavaara M, Nevalainen A, Martelius T, Kuusela P, Anttila VJ.** Impact of short-incubation MALDI-TOF MS on empiric antibiotic therapy in bloodstream infections caused by Pseudomonas aeruginosa, Enterococcus spp. and AmpC-producing Enterobacteriaceae. *Diagn Microbiol Infect Dis. 2019;94:1-6.*

51. **Curtoni A, Cipriani R, Marra ES, Barbui AM, Cavallo R, Costa C.** Rapid identification of microorganisms from positive blood culture by MALDI-TOF MS after short-term incubation on solid medium. *Curr Microbiol. 2017;74:97-102.*

52. **Kock R, Wullenweber J, Horn D, Lanckohr C, Becker K, Idelevich EA.** Implementation of short incubation MALDI-TOF MS identification from positive blood cultures in routine diagnostics and effects on empiric antimicrobial therapy.

Antimicrob Resist Infect Control. 2017;6:12.

53. **Sauget M, Bertrand X, Hocquet D.** Rapid antibiotic susceptibility testing on blood cultures using MALDI-TOF MS. *PLoS One. 2018;13:e0205603.*

54. **Anantharajah A, Tossens B, Olive N, Kabamba-Mukadi B, Rodriguez-Villalobos H, Verroken A.** Performance evaluation of the MBT STAR® - Carba IVD assay for the detection of carbapenemases with MALDI-TOF MS. *Front Microbiol. 2019;10:1413.*

55. **Axelsson C, Rehnstam-Holm AS, Nilson B.** Rapid detection of antibiotic resistance in positive blood cultures by MALDI-TOF MS and an automated and optimized MBT-ASTRA protocol for Escherichia coli and Klebsiella pneumoniae. *Infect Dis. 2020;52:45-53.*

56. **Yu J, Liu J, Li Y, Yu J, Zhu W, Liu Y, et al.** Rapid detection of carbapenemase activity of Enterobacteriaceae isolated from positive blood cultures by MALDI-TOF MS. *Ann Clin Microbiol Antimicrob. 2018;17:22.*

57. **Nix ID, Idelevich EA, Storck LM, Sparbier K, Drews O, Kostrzewa M, et al.** Detection of methicillin resistance in staphylococcus aureus from agar cultures and directly from positive blood cultures using MALDI-TOF mass spectrometry-based direct-on-target microdroplet growth assay. *Front Microbiol. 2020;11:232.*

58. **Idelevich EA, Storck LM, Sparbier K, Drews O, Kostrzewa M, Becker K.** Rapid direct susceptibility testing from positive blood cultures by the matrix-assisted laser desorption ionization-time of flight mass spectrometry-based direct-on-target microdroplet growth assay. *J Clin Microbiol. 2018;56:e00913-18.*

59. **Kim J, Wang H young, Kim S, Park SD, Yu K, Kim HY, et al.** Evaluation of the Punch-itTM NA-Sample kit for detecting microbial DNA in blood culture bottles using PCR-reverse blot hybridization assay. *J Microbiol Methods. 2016;128:24-30.*

60. **Dubourg G, Raoult D, Fenollar F.** Emerging methodologies for pathogen identification in bloodstream infections: an update. *Expert Rev Mol Diagn. 2019;19:161-73.*

61. **Hazelton BJ, Thomas LC, Unver T, Iredell JR.** Rapid identification of Gram-positive pathogens and their resistance genes from positive blood culture broth using a multiplex tandem RT-PCR assay. *J Med Microbiol. 2013;62:223-31.*

62. **Rocchetti TT, Martins KB, Martins PYF, Oliveira RA de, Mondelli AL, Fortaleza CM, et al.** Detection of the mec A gene and identification of Staphylococcus directly from blood culture bottles by multiplex polymerase chain reaction. *Braz J Infect Dis. 2018;22:99-105.*

63. **Shanmugakani RK, Fujiya Y, Akeda Y, Hamaguchi S, Hamada S, Tomono K.** Rapid multiplex detection of the resistance genes mecA, vanA and vanB from Gram-positive cocci-positive blood cultures using a PCR- dipstick technique. *J Med Microbiol. 2020;69:249-55.*

64. **Berinson B, Both A, Berneking L, Christner M, Lutgehetmann M, Aepfelbacher M, et al.** Usefulness of BioFire FilmArray BCID2 for blood culture processing in clinical practice. *J Clin Microbiol. 2021;59:e0054321.*

65. **Ray STJ, Drew RJ, Hardiman F, Pizer B, Riordan A.** Rapid identification of microorganisms by FilmArray blood culture identification panel improves clinical management in children. *Pediatr Infect Dis J. 2016;35:e134-8.*
66. **Salimnia H, Fairfax MR, Lephart PR, Schreckenberger P, DesJarlais SM, Johnson JK, et al.** Evaluation of the FilmArray blood culture identification panel: results of a multicenter controlled trial. *J Clin Microbiol. 2016;54687-98.*
67. **Verhoeven PO, Haddar CH, Rigaill J, Fonsale N, Carricajo A, Grattard F, et al.** Comparison of the Fully Automated FilmArray BCID Assay to a 4-Hour culture test coupled to mass spectrometry for day 0 identification of microorganisms in positive blood cultures. *BioMed Res Int. 2018;2018:1-6.*
68. **Payne M, Champagne S, Lowe C, Leung V, Hinch M, Romney MG.** Evaluation of the FilmArray Blood Culture Identification Panel compared to direct MALDI-TOF MS identification for rapid identification of pathogens. *J Med Microbiol. 2018;67:1253-6.*
69. **Verroken A, Despas N, Rodriguez-Villalobos H, Laterre PF.** The impact of a rapid molecular identification test on positive blood cultures from critically ill with bacteremia: A pre-post intervention study. *PLoS One. 2019;14:e0223122.*
70. **Fhooblall M, Nkwanyana F, Mlisana KP.** Evaluation of the BioFire® FilmArray® Blood Culture Identification Panel on positive blood cultures in a regional hospital laboratory in KwaZulu-Natal. *Afr J Lab Med. 2016;5:411.*
71. **Pardo J, Klinker KP, Borgert SJ, Butler BM, Giglio PG, Rand KH.** Clinical and economic impact of antimicrobial stewardship interventions with the FilmArray Blood Culture Identification Panel. *Diagn Microbiol Infect Dis. 2016;84:159-64.*
72. **Saito K, Endo S, Katsumi M, Ishizawa C, Fujikawa Y, Inomata S, et al.** Evaluation of the FilmArray Blood Culture Identification Panel on detection of pathogenic microorganisms in positive blood cultures: the first clinical report in Japan. *Jpn J Infect Dis. 2018;71:145-7.*
73. **Huang TD, Melnik E, Bogaerts P, Evrard S, Glupczynski Y.** Evaluation of the ePlex Blood Culture Identification Panels for detection of pathogens in bloodstream infections. *J Clin Microbiol. 2019;57:e01597-18.*
74. **Schmitz JE, Tang YW.**The GenMark ePlex ®: another weapon in the syndromic arsenal for infection diagnosis. *Future Microbiol. 2018;13:1697-708.*
75. **Carroll KC, Reid JL, Thornberg A, Whitfield NN, Trainor D, Lewis S, et al.** Clinical performance of the novel GenMark Dx ePlex Blood Culture ID Gram-Positive Panel. *J Clin Microbiol. 2020;58:e01730-19.*
76. **Wolk DM, Young S, Whitfield NN, Reid JL, Thornberg A, Carroll KC, et al.** A multicenter clinical study to demonstrate the diagnostic accuracy of the GenMark Dx ePlex Blood Culture Identification Gram-Negative Panel. *J Clin Microbiol. 2021;59:e02484-20.*
77. **Guevara R, Davis MG, Davis MA, Ahmed S, Shah S, Cuaron K, et al.** Efficacy of VERIGENE® blood culture assays in accurately and rapidly detecting Gram-negative and Gram-positive pathogens. *Infect Disord Drug Targets.*

2022;22:e100122200121.

78. **Nazar S.** Implementation Plan to Integrate the Nanosphere Verigene Assay in a High-volume Reference Laboratory and the Anticipated Workflow Benefits to Adopting Rapid Blood Culture Testing [Thesis]. *San Rafael: Dominican University of California; 2018.*

79. **Farkas J.** PulmCrit- Nucleic acid microarrays streamline bacteremia management. *[Online]. 2019 [Accessed 21/08/2022], Available from URL:https://emcrit.org/pulmcrit/verigene/*

80. **Beckman M, Washam MC, DeBurger B, Haslam DB, Courter JD, Andersen H, etal.** Reliability of the Verigene system for the identification for Gram-positive Bacteria and detection of antimicrobial resistance markers from children with bacteremia. *Diagn Microbiol Infect Dis. 2019;93:191-5.*

81. **Hayakawa K, Mezaki K, Kobayakawa M, Yamamoto K, Mutoh Y, Tsuboi M, et al.** Impact of rapid identification of positive blood cultures using the Verigene system on antibiotic prescriptions: A prospective study of community-onset bacteremia in a tertiary hospital in Japan. *PLoS One. 2017;12:e0181548.*

82. **Freiberg JA, Deri CR, Nesbitt WJ, Humphries RM, Nelson GE.** Identification and antibiotic susceptibility patterns of clinical blood culture isolates not identified by a rapid microarray diagnostic system. *Microbiol Spectr. 2021;9:e00175-21.*

83. **Burrack-Lange SC, Personne Y, Huber M, Winkler E, Weile J, Knabbe C, et al.** Multicenter assessment of the rapid Unyvero Blood Culture molecular assay. *J Med Microbiol. 2018;67:1294-301.*

84. **Ashikawa S, Tarumoto N, Imai K, Sakai J, Kodana M, Kawamura T, et al.** Rapid identification of pathogens from positive blood culture bottles with the MinION nanopore sequencer. *J Med Microbiol. 2018;67:1589-95.*

85. **Prudent E, Raoult D.** Fluorescence in situ hybridization, a complementary molecular tool for the clinical diagnosis of infectious diseases by intracellular and fastidious bacteria. *FEMS Microbiol Rev. 2019;43:88-107.*

86. **Frickmann H, Zautner AE, Moter A, Kikhney J, Hagen RM, Stender H, et al.** Fluorescence in situ hybridization (FISH) in the microbiological diagnostic routine laboratory: a review. *Crit Rev Microbiol. 2017;43:263-93.*

87. **Zrodlowski T, Sobonska J, Salamon D, McFarlane IM, Zi^tkiewicz M, Gosiewski T.** Classical microbiological diagnostics of bacteremia: are the negative results really negative? What is the laboratory result telling us about the "gold standard"? *Microorganisms. 2020;8:346.*

88. **Zhang B, Maimaiti Y, Liu C, Li J, Wang H, Lin H, et al.** Direct detection of Staphylococcus aureus in positive blood cultures through molecular beacon-based fluorescence in situ hybridization. *J Microbiol Methods. 2019;159:34-41.*

89. **Koncelik DL, Hernandez J.** The impact of implementation of rapid quick FISH testing for detection of coagulase-negative staphylococci at a community-based hospital: Table 1. *Am J Clin Pathol. 2016;145:69-74.*

90. **Enroth H, Retz K, Andersson S, Andersson C, Svensson K, Ljungstrom L, et**

al. Evaluation of QuickFISH and maldi Sepsityper for identification of bacteria in bloodstream infection. *Infect Dis. 2019;51:249-58.*

91. **Cenci E, Paggi R, Socio GVD, Bozza S, Camilloni B, Pietrella D, et al.** Accelerate Pheno™ blood culture detection system: a literature review. *Future Microbiol. 2020;15:1595-605.*

92. **Starr KF, Robinson DC, Hazen KC.** Performance of the accelerate diagnostics Pheno™ system with resin-containing BacT/ALERT® Plus blood culture bottles. *Diagn Microbiol Infect Dis. 2019;94:122-8.*

93. **Descours G, Desmurs L, Hoang TL, Ibranosyan M, Baume M, Ranc AG, et al.** Evaluation of the Accelerate Pheno™ system for rapid identification and antimicrobial susceptibility testing of Gram-negative bacteria in bloodstream infections. *Eur J Clin Microbiol Infect Dis. 2018;37:1573-83.*

94. **Rodel J, Bohnert JA, Stoll S, Wassill L, Edel B, Karrasch M, et al.** Evaluation of loop-mediated isothermal amplification for the rapid identification of bacteria and resistance determinants in positive blood cultures. *Eur J Clin Microbiol Infect Dis. 2017;36:1033-40.*

95. **Heraud S, Freydiere AM, Doleans-Jordheim A, Bes M, Tristan A, Vandenesch F, et al.** Direct identification of *Staphylococcus aureus* and determination of methicillin susceptibility from positive blood-culture bottles in a Bact/ALERT system using Binax Now S. aureus and PBP2a Tests. *Ann Lab Med. 2015;35:454-7.*

96. **Berendsen EM, Levin E, Braakman R, Prodan A, van Leeuwen HC, Paauw A.** Untargeted accurate identification of highly pathogenic bacteria directly from blood culture flasks. *Int J Med Microbiol. 2020;310:151376.*

97. **Drees C, Vautz W, Liedtke S, Rosin C, Althoff K, Lippmann M, et al.** GC-IMS headspace analyses allow early recognition of bacterial growth and rapid pathogen differentiation in standard blood cultures. *Appl Microbiol Biotechnol. 2019;103:9091-101.*

98. **Altun O, Athlin S, Almuhayawi M, Stralin K, Ozenci V.** Rapid identification of *Streptococcus pneumoniae* in blood cultures by using the ImmuLex, Slidex and Wellcogen latex agglutination tests and the BinaxNOW antigen test. *Eur J Clin Microbiol Infect Dis. 2016;35:579-85.*

99. **Le Page S, Raoult D, Rolain JM.** Real-time video imaging as a new and rapid tool for antibiotic susceptibility testing by the disc diffusion method: A paradigm for evaluating resistance to imipenem and identifying extended-spectrum в-lactamases. *Int JAntimicrob Agents. 2015;45:61-5.*

100. **Costa-de-Oliveira S, Teixeira-Santos R, Silva AP, Pinho E, Mergulhao P, Silva-Dias A, et al.** Potential impact of flow cytometry antimicrobial susceptibility testing on the clinical management of Gram-negative bacteremia using the FASTinov® Kit. *Front Microbiol. 2017;8:2455.*

101. **Athamna A, Freimann S.** A new rapid method for detecting extended- spectrum beta-lactamase/AmpC-producing Enterobacteriaceae directly from positive blood

cultures using the Uro4 HB&LTM system. *Braz J Microbiol. 2019;50:927-33.*

102. **Kim HS, Kim JO, Lee JE, Park KG, Lee HK, Kim SY, et al.** Performance of a novel fluorogenic assay for detection of carbapenemase- producing enterobacteriaceae from bacterial colonies and directly from positive blood cultures. *J Clin Microbiol. 2019;58:e01026-19.*

103. **Rosselin M, Prod'hom G, Greub G, Croxatto A.** Performance evaluation of the Quantamatrix QMAC-dRAST system for rapid antibiotic susceptibility testing directly from blood cultures. *Microorganisms. 2022;10:1212.*

104. **Choi J, Jeong HY, Lee GY, Han S, Han S, Jin B, et al.** Direct, rapid antimicrobial susceptibility test from positive blood cultures based on microscopic imaging analysis. *Sci Rep. 2017;7:1148.*

105. **Falconer K, Hammond R, Parcell B, Gillespie SH.** Rapid determination of antimicrobial susceptibility of gram-negative bacteria from positive blood cultures using an innovative method. *[Online]. 2020 [Accessed 21/08/2022], available at URL:https://www. microbiologyresearch. org/content/journal/acmi/10.1099/acmi.fis2019.po0072*

106. **Seco BM, Campos JC, da Costa Rocha DA, de Lima AV, de Oliveira FF, Lemo MEB, et al.** Improved blood culture workflow for faster identification of KPC-producing Enterobacterales. *Braz J Microbiol. 2019;50:127-32.*

107. **Grinberg S, Schubert S, Hochauf-Stange K, Dalpke AH, Narvaez Encalada M.** Saving time in blood culture diagnostics: a prospective evaluation of the qvella FAST-PBC prep application on the fast system. *J Clin Microbiol. 2022;60:e0253321.*

108. **Kuo P, LeCrone K, Chiu M, Realegeno S, Pride DT.** Analysis of the FASTTM System for expedited identification and antimicrobial susceptibility testing of bloodborne pathogens. *Diagn Microbiol Infect Dis. 2022;115783.*

MIX
Papier aus verantwortungsvollen Quellen
Paper from responsible sources
FSC® C105338

Printed by Books on Demand GmbH, Norderstedt / Germany